Pharmaceutical Contamination Control:
Practical Strategies for Compliance

Nigel Halls
Editor

PDA
Bethesda, MD, USA
DHI Publishing, LLC
River Grove, IL, USA

10 9 8 7 6 5 4 3 2 1

ISBN: 1-933722-02-9

PDA
4350 East West Highway
Suite 200
Bethesda, MD 20814
United States
301-986-0293

Davis Healthcare International Publishing, LLC
2636 West Street
River Grove
IL 60171
United States
www.DHIBooks.com

CONTENTS

1

PHARMACOPOEIAL WATER SYSTEMS AND THE RISKS OF MICROBIOLOGICAL CONTAMINATION FROM WATER

Nigel Halls

Giant strides have been made with respect to minimising the risks of microbiological contamination of pharmaceutical dosage from manufacture by air-borne microorganisms, such that, in aseptic manufacture at least, the greatest source of contamination is now considered to be from personnel.

Contamination from air and personnel is mainly from gram-positive bacteria.

Water is the principle source of gram-negative bacteria. If present in pharmaceutical dosage forms, gram-negative bacteria present greater risks of infection to the patient and more likelihood of deleterious chemical changes to the product than would be presented by gram-positive bacteria.

This monograph addresses water from the perspective of its potential to microbiologically contaminate dosage forms during their manufacture and storage in pharmaceutical facilities.

WATER AS A HABITAT FOR MICROORGANISMS

Water is one of the most commonly found substances on this planet, but 'pure' water in the sense of H_2O uncontaminated by dissolved and/or suspended substances is very rare. As the concentration of dissolved and suspended contaminants increases in water so the terminology changes. It stops being water and starts becoming a solution, a suspension, a broth, a soup, a stew, a jelly, etc. For water to be termed water it must contain only a very low concentration of contaminants of any type.

Pure water should contain no organic nutrients at all. Such a quality of water can never be practically obtained, but nonetheless water which remains recognisable as water can contain enough organic nutrients to support microbial growth.

Some microorganisms have evolved to be able to survive and grow in environments where there are only very low concentrations of organic nutrients. This evolution entails two survival mechanisms. First, they have evolved a means of surviving with minimal wastage of energy. Second, they have evolved a level of metabolic versatility sufficient to utilise almost any type of complex molecule as an energy source, because the choice may be between whatever is available or starvation and death.

The key evolutionary feature of gram-negative bacteria which allows survival and growth in water is in their cell wall or cell envelope.

The structural rigidity of the bacterial cell wall is conferred by a material called peptidoglycan. In gram-positive bacteria this is present as a thick layer which is outermost in the cell wall. In gram-negative bacteria, the peptidoglycan is only a thin layer and not the outermost layer. Gram-negative bacteria are sometimes described as having a cell envelope rather than a cell wall, a loosely attached layer of material called lipo-polysaccharide located outside a thin structural layer of peptidoglycan.

The important evolutionary advantages conferred by lipopolysaccharide are:

- it contributes to adhesion to surfaces of gram-negative bacteria allowing them to form as biofilms in aqueous environments
- it 'attracts' and 'entraps' organic macromolecules from aqueous environments
- it allows for entrapped organic macromolecules to be 'recognised' by the cell so that specific enzymes can be synthesised and broken down into smaller fragments capable of passing through the peptidoglycan layer into the cell
- it retains the enzymes synthesised by the cell so they are not lost into the external environment.

The ability of gram-negative bacteria to form as biofilms (Donlan, 2002; Dunne, 2002; Watnick & Kolter, 2000) on surfaces over which water flows is the primary mechanism for survival in a low nutrient environment. Nutrients come to the bacteria instead of the bacteria using energy to find nutrients.

Biofilms form to some extent on all surfaces which come into contact with stagnant or slow flowing water. Initially, biofilm-forming bacteria present in water as free-floating forms, are attracted by one or other of various physical and chemical mechanisms to surfaces where attachment takes place.

Once attached, the bacteria develop capsular material around themselves and, given sufficient nutrients, develop as a monolayer biofilm. As time goes by the film thickens as a result of the bacteria dividing in more than one plane (Anon, 1993; Stoodley et al, 2001; Hunt et al, 2004). As gram-negative biofilm-forming bacteria rely on oxidative metabolism, the thickness of the biofilm becomes limited by the increasing inability of sufficient oxygen to penetrate to those bacteria which initially adhered to the surface. Without oxygen these bacteria die off, they lyse and release chemically active metabolic by-products which disrupt their adhesion to the surface. The biofilm breaks away, or 'sloughs off' into the water course; potentially the whole cycle — beginning with attraction and ending with sloughing off — then begins again.

The types of bacteria which form as biofilm are generally either pseudomonads or identifiable with the genus *Acinetobacter*. The term 'pseudomonad' refers to bacteria which resemble those identifiable with the genus *Pseudomonas*. Formerly this was a genus containing a very large number of species; but with the development of technologies over the last decades of the 20th century for better analysing the genome, it has been 'split' into several other genera which appear to have evolved separately. These include *Brevundimonas*, *Burkholderia*, *Ralstonia*, *Stenotrophomonas*, etc.

It should be recognised that water is not necessarily an inimical environment to microorganisms which do not form biofilm. Many microorganisms recovered from natural waters are not intrinsic to water. Some microorganisms can be found as 'free-floating' types, but because of the very low concentration of nutrients in water it is extremely rare for these 'free-floating' types to do more than just survive in water without metabolising, growing or increasing in numbers. Organisms may be present in water which are intrinsic to soil (for instance, the spore-forming genus *Bacillus*) but have been washed into water courses by rainfall. Other microorganisms may have been passed into water courses from animal excreta (*Salmonella*, for example).

COMPENDIAL STANDARDS FOR MICROBIOLOGICAL QUALITY OF WATER

Water is used extensively in pharmaceutical manufacture — as an ingredient, for cleaning and as a diluent for various purposes. The extent to which it is unavoidable in manufacturing facilities makes it, unless carefully controlled, a very likely cause of compromise to the chemical, physical and biological purity which is expected of pharmaceutical products.

For this reason compendial limits are placed on the quality of water which may be used in the manufacture of pharmaceuticals.

The way in which these limits are presented differ between *PhEur* and *USP*.

Three monographs deal with water in *PhEur* — for *Water, Purified (Aqua purificata)*, for *Water for Injections (Aqua ad injectabilia)*, and for *Water, Highly Purified (Aqua valde purificata)*.

Within the monograph for *Water, Purified*, *PhEur* distinguishes *Purified Water in Bulk* from *Purified Water in Containers* to accommodate the differences between water used as a pharmaceutical 'product' and water used in manufacture. Similarly, the *PhEur* monograph for *Water for Injections* distinguishes *Water for Injections in Bulk* from *Sterilised Water for Injections*.

However, the monograph on *Highly Purified Water* is not split in this way. *Highly Purified Water* is described in the *PhEur* monograph as intended 'for use in the preparation of medicinal products'.

USP has produced monographs for *Purified Water*, *Water for Injection*, and for various other applications of water as a pharmaceutical 'product'. In relation to the use of water in manufacture, the two monographs are supported by a general chapter, *<1231> Water for Pharmaceutical Purposes*.

The limits placed on water by the pharmacopoeiae are largely harmonised.

- The limits placed on *Water, Purified* (*PhEur*) do not differ significantly from those placed on *Purified Water* (*USP*).

- The limits placed on *Water for Injections* (*PhEur*) do not differ significantly from those placed on *Water for Injection* (*USP*).

The *PhEur* limits for *Water for Injections* and for *Highly Purified Water* are identical. The difference between the two types of water is only in the methods allowable for their preparation (*Water for Injections, PhEur* may only be prepared by distillation).

Two types of microbiological limits for water are applied by the pharmacopoeiae:

- Quantitative limits on concentrations of aerobic microorganisms, described as total viable aerobic counts in *PhEur* and defined in the monographs for *Purified Water in Bulk* and for *Water for Injections in Bulk*. Microbial limits are not included in the *USP* monographs for water, but are defined in *<1231> Water for Pharmaceutical Purposes* where they are described as 'action limits'.

    — The microbiological limits applying to concentrations of microorganisms recoverable from *Purified Water* (*USP*) and from *Water, Purified* (*PhEur*) are for there to be not more than 100 cfu/mL.

    — The microbiological limits applying to concentrations of microorganisms recoverable from *Water for Injection* (*USP*) and from *Water for Injections* (*PhEur*) are for there to be not more than 10 cfu/100 mL.

- Quantitative limits on concentrations of endotoxin of not more than 0.25 EU/mL are contained in the monographs for *Water for Injection* (*USP*), *Water for Injections* (*PhEur*) and *Highly Purified Water* (*PhEur*).

There is a fundamental distinction between the compendial limits placed on concentrations of aerobic microorganisms in water and those placed on endotoxin concentrations.

The endotoxin limit is a true 'pass–fail' or 'accept–reject' limit. Water cannot be identifiable as *Water for Injection* (*USP*), *Water for Injections* (*PhEur*) or *Highly Purified Water* (*PhEur*) unless this limit is complied with.

On the other hand both *USP* and *PhEur* emphasise that batch 'pass–fail' decision-making is not practical for microbiological limits applying to pharmacopoeial grade waters 'in bulk'. This is for two reasons. First, because in bulk production, water is not generally 'batched', it is prepared continuously or semi-continuously and usually kept in continuous motion and with continuous mixing. Second, because there is generally at least a 48-hour delay between sampling and obtaining data from microbiological testing of water, by which time the water represented by the sample will probably have been diluted in water prepared both earlier and later, and have already been used in production or for cleaning.

To get round these impracticalities, the compendia identify the quantitative limits on concentrations of microorganisms as 'action limits'. The idea of action limits is that when an excursion beyond them arises it should not require 'rejection' of the water in question, or of material manufactured using the water, or of material washed with the water. Instead, excursions should require some corrective action in the preparation (or storage and distribution) process to bring it back within limits.

In addition to 'action limits' which according to *USP*, indicate that 'a process has drifted from its normal operating range', both pharmacopoeiae mention 'alert limits'. These provide a warning of a possible deleterious trend away from the normal operating condition of the process. *USP* states that infringement of an 'alert limit' while still in compliance with an 'action limit' does 'not necessarily require a corrective action.'

These compendial principles merit some more detailed consideration.

Microbiological Limits for Purified Water

The action limits recommended by the pharmacopoeiae for purified water are that there should be not more than 100 cfu/mL.

Recovery of microorganisms from any potential habitat is a function of the media and incubation conditions used. This is true for water as for any other habitat. It has long been known that from nutritionally depleted habitats such as water in which gram-negative bacteria predominate, higher numbers are recoverable using nutritionally depleted media, low incubation temperatures and longer incubation periods (Topping, 1937; van der Linde et al, 1999; Punakabutra et al, 2004).

USP <1231> refers to media with high-nutrient content incubated at 30–35°C for 48–72 hours as 'classical' methodology for counting microorganisms in water. Decisions as to exactly what methodology to apply are left to the user of the water. *PhEur* mandates the use of a low-nutrient medium (Medium R2A — Reasoner & Geldreich, 1985) and longer incubation (five days) for water testing, but does not go all the way with lower temperatures (the recommended incubation temperature remains 30–35°C).

Although somewhat higher numbers may be expected, the downside of the *PhEur* approach is that the delay in obtaining the data is longer, and therefore a 'drift' out of the normal operating range may have become persistent before it has even been recognised.

As a means of monitoring changes in the process, it is therefore more meaningful to ensure that the means of recovery is consistently applied than for the method of recovery to be necessarily optimised.

It is illuminating to compare statements made 40 to 50 years ago by the UK government standard applying to water in general.

In 1969, the UK standard for bacteriological examination of water supplies (Anon, 1969) states:

> 'Colony counts provide an estimate of general bacterial purity, which is of particular value when water is used industrially ... These counts do not represent the total number of bacteria present but merely the number of bacterial cells or groups of bacterial cells capable of producing visible colonies in the medium used within the given times and at the particular temperature ... Colony counts are particularly susceptible to variations in technique and to errors in sampling. Since much value lies in the comparison of repeated samples from the same source, it is important to adhere to a standard technical method ...'

For pharmacopoeial waters it should be realised that the pharmacopoeial limit for *Purified Water* of nmt 100 cfu/mL is not a reflection of the process capability of current technologies. It would be extremely rare to find pharmaceutical water systems, validated properly and operating under control, to return typical total aerobic viable counts of more than about 20 cfu/mL, with many regularly returning counts of less than 10 cfu/mL. It is extremely unlikely therefore that the choice of method used (*USP* 'classical' or *PhEur*) is going to have a significant influence on the number or frequency of excursions beyond the recommended action limit.

In practical terms, most modern *Purified Water* systems could drift significantly out of the 'process normal operating condition' without exceeding the action limit of 100 cfu/mL. Effectively this means that for sensible control a limit much lower than 100 cfu/mL must be established against which corrective action should be taken when excursions arise. This is termed an 'alert' limit within the pharmacopoeial definition (*USP <1231>*) but according to the pharmacopoeiae excursions beyond alert limits do not require any corrective action to be taken.

The practicality of establishing process-based alert limits is something which can only be cautiously approached from a general standpoint. A sensible approach is to first gather data and determine the manner or 'shape' in which it is distributed. This is done by the frequency histogram method in which the numbers (or percentages) of zero counts, counts of one colony, counts of two colonies etc, per sample are plotted on a graph. If the graph has a regular shape it may be possible

Figure 1.1 Frequency histogram of microbiological data from a purified water system. Typically for pharmaceutical water systems these microbiological data do not fit a Gaussian (Normal) Distribution. The pharmacopoeial limit of not more than 100 cfu/mL is however inappropriately high

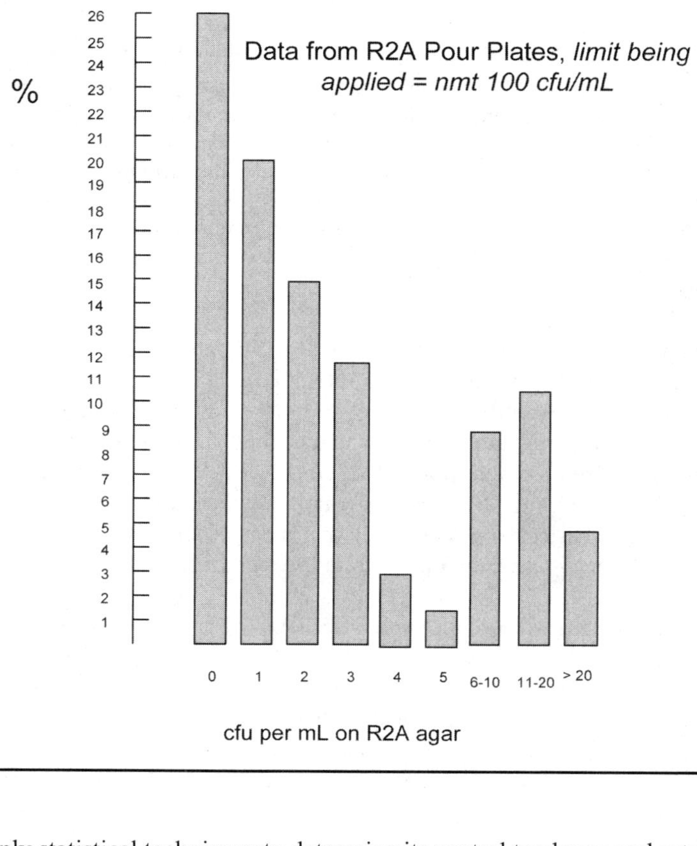

to apply statistical techniques to determine its central tendency and establish limits on the probability of exceeding particular numbers.

Given that the average count per sample for *Purified Water* may be greater than 5 cfu, and the expected range may be up to 10 or 20 cfu per sample, the numbers allow for the possibility of establishing alert limits lower than the pharmacopoeial action limit for *Purified Water*.

However two critical questions should be addressed in relation to alert limits:

- First, what is to be achieved by knowing that an alert limit has been exceeded while taking no corrective action?

  In some situations (eg counter-terrorism), increased vigilance can be effective in postponing or preventing problems, but it is difficult to see how this can apply to the microbiological quality of pharmaceutical water systems.

- Second, is it to be really expected that the microbiological quality of pharmaceutical water systems drifts progressively out of control, returning increasingly higher data until alert and then action limits are exceeded?

  This is in fact unlikely, both from empirical experience of handling microbiological data and from understanding how biofilm builds up and sloughs off in a pattern of repeating cycles. Excursions beyond action limits rarely arise after a series of warnings in which alert limits are exceeded. High counts arise as random events or 'spikes', which are independent of the general patterns of data distribution. This is perhaps not surprising when considering how a biofilm slough can create a sudden and unpredictable increase in counts well beyond that 'normally' seen in water samples.

So in relation to alert limits, 'received wisdom' is that they represent a sensible and cautious approach to control. In practice, they may not be applicable to microbiological control of *Purified Water* systems and, if they result in 'false alarm' after 'false alarm', may indeed be counter-productive to good microbiological control. At the same time it seems that critical questioning of their value is so rare that they have become a regulatory expectation, so challenges to them at inspection should, in the interests of 'the business' be taken with extreme caution.

It may be sensible to re-label alert limits as 'investigational'. In some instances actions may result from investigation, in others the outcome of the investigation may require no action. A third option is that the investigation may be inconclusive.

Microbiological Limits for Water for Injection, Water for Injections and Highly Purified Water

The action limits recommended by the pharmacopoeiae for *Water for Injection*, *Water for Injections* and *Highly Purified Water* are that there should be not more than 10 cfu/100 mL.

PhEur adds that testing should be done by the membrane filtration method (commonly used for all pharmacopoeial grade water testing) and that the volume filtered should be at least 200 mL. The restriction on the volume tested follows a 'rule of thumb' which allows that the sample size for a test applied to discrete variables should be such that if the sample were at its limit, there would be at least 20 units (in this case colonies) countable.

Water for Injections PhEur must be prepared by distillation. *Highly Purified Water PhEur* can effectively be prepared by any method which can achieve its specification (the same specification as *Water for Injections PhEur*). *Water for Injection USP* can be prepared by either distillation or reverse osmosis (RO).

Pharmacopoeial grades of water are hot when prepared by distillation. In addition, good manufacturing practice (GMP) requires that *Water for Injection(s)* is stored and distributed in conditions designed to prevent growth of microorganisms or any other contamination. Within GMP these two factors have combined into a general requirement that a temperature in excess of 75°C should be maintained in all parts of all storage and distribution systems used in connection with *Water for Injection(s)*.

Gram-negative bacteria do not survive and biofilm does not form at temperatures much higher than 60–65°C. Gram-positive vegetative bacteria do not survive at temperatures much higher than 75°C, and even gram-positive spore-forming bacteria which may survive at these temperatures are not capable of growing, dividing and increasing in numbers.

As a result of high temperature storage and circulation, the typical microbiological counts from *Water for Injection(s)* systems are zero colonies per 200 mL sample. At least one laboratory known to the author has, over several years, been routinely testing water samples taken from a high temperature circulating *Water for Injection(s)* system by a pharmacopoeial *Test for Sterility* and has been ordinarily in compliance with its requirements. This approach to water testing is, however, not recommended. It is not a requirement. A potentially quantitative test option is available and normally used, and it introduces risks of contamination in the laboratory and of contaminating product *Tests for Sterility* carried out in the same time frame.

The pattern of zero count data obtained from *Water for Injection(s)* leaves little leeway for establishing either action or alert limits. It is probably sensible to consider any recovery of gram-negative bacteria (even one colony) from a high temperature re-circulating *Water for Injection(s)* system as an action limit, and any recovery of gram-positive bacteria or moulds as an alert limit. Some laboratories set limits according to the frequency of repeating recoveries from *Water for Injection(s)* systems, eg an action is required if there is more than one count in five, 10 or 20 successive samples. This further dilutes the possibility of an investigation, and of some previously unknown 'problem', lapse in control, or change in engineering practice coming to management's attention — again this practice cannot be recommended.

Where *Water for Injection* is prepared by RO as allowed by *USP*, or where *Highly Purified Water PhEur* is prepared by RO and in any other ambient temperature system, the approach to establishing alert and action limits should be similar to the one recommended above for *Purified Water* — plot the actual data on a frequency histogram, and determine any regularity to its shape. Limits may be set by consulting a statistician regarding the best approach, or more probably, may be set arbitrarily. It is extremely unlikely that the microbiological and endotoxin limits placed on *Water for Injection(s)* and *Highly Purified Water* can be consistently maintained in ambient temperature storage and distribution systems because of the potential for formation of biofilm.

PHARMACOPOEIAL WATER SYSTEMS

The design of pharmacopeial water systems is a specialised and skilled engineering discipline. Without wishing to appear to minimise other aspects of water system engineering design, it is probably fair to say that the uniqueness of pharmacopoeial water systems lies in its focus on microbiological control.

Only very rarely are there difficulties in achieving the physical and chemical limits required of pharmacopoeial waters. On the other hand, microbiological problems originating from biofilm occur in most water systems from time to time.

The primary focus in the design, construction and operation of pharmacopoeial water systems is on the prevention of initial biofilm formation (adhesion). Biofilm forms most easily when water is stagnant or flows only slowly over surfaces, and when the surfaces over which it flows are rough or uneven. Conversely, when water flows rapidly over surfaces, the shear forces created by its movement act against the forces which attract microorganisms to the surfaces and act against adhesion. Where the surfaces are smoothest there are fewer opportunities for adhering microorganisms to be protected from these shear forces.

Therefore the primary microbiological control principles applied in pharmacopoeial water systems are:

- *Constant flow*. Biofilm will form in parts of water systems where water is stagnant or where flow rates are low. Pharmacopoeial water systems are generally designed to have flow rates of 1–3 m per sec at the slowest points in their distribution loops. The origins of these flow rate parameters are obscure and probably derive more from experience than science. It is not impossible to obtain satisfactory in systems which do not operate within these flow rates.

- *Smooth surfaces*. Biofilm will form in parts of water systems where there are irregular and rough surfaces. The measure of smoothness of engineered surfaces is roughness average (Ra µm) (roughness average — a measure obtained from an instrument which drags a profiled diamond over about 0.8mm of surface area), and the internal finish for most pipe-work in pharmacopoeial water systems is specified to be around 0.5 µm Ra.

Other secondary microbiological controls are incorporated in all pharmacopoeial water systems, for instance ultraviolet (UV) lights and/or high temperature or chemical sanitisation procedures. These acknowledge the extraordinary difficulty of designing and building practical systems for preparing and delivering water of the required quality, in the required quantities, to the required locations, at the required times without incurring some features which risk biofilm formation.

All pharmacopoeial water systems comprise three separate stages:

- *Pre-treatment*: processes designed to minimise the variability of the incoming feed-water.

- *Preparation*: the process or processes whereby impurities are removed from the water to pharmacopoeial standards.

- *Storage and distribution*: the means by which the pharmacopoeial grade water is delivered to its use points without compromising its quality.

Biofilm formation which could ultimately result in compromising the microbiological quality of the pharmacopoeial grade water can potentially arise in any one or all of these stages.

Pre-treatment

All processes are affected by the 'rubbish in — rubbish out' principle.

The feed water is at the beginning of all water systems. Aside from the GMP requirement that feed water supplied to pharmacopoeial water systems should be of potable quality (suitable for drinking and dietetic purposes) there is an enormous potential for variability among, and even within, feed waters according to climatic conditions and time of year. Pre-treatments are intended to minimise this variability, ensuring that a consistent 'raw material' is presented to the preparation stage.

Sand filters, softeners, chlorination, and carbon filters are all typical pre-treatments. They all may be seen in some water systems; but only some in others.

- *Sand filters* — are used to remove large organic particles, insects, etc.

- *Softeners* — are used to remove 'hardness' from water thus preventing internal surfaces of pipe-work and equipment becoming clogged by 'limescale' or 'fur'.

- *Chlorination*— many pharmaceutical companies choose to chlorinate or re-chlorinate the feed water to their water systems. If the feed water comes from a municipal supply it may already have been chlorinated. Bacteriological safety, in the sense of freedom from intestinal pathogens rather than control of overall microbial numbers, is the main reason for chlorination of municipal potable water supplies. The concentration of non-pathogenic microorganisms in these supplies may still vary considerably as a function of the water source (well water or surface water), the distances and condition of the pipe-work through which the water is pumped, and the time of year.

- *Carbon filters* — are used, paradoxically, to remove the additional chlorine added in re-chlorination. They are also of some value in removing colour — some natural waters have an unacceptable yellow tinge resulting from percolation through 'peaty' sub-strata.

With the exception of chlorination, all these types of pre-treatment have the potential to allow the formation of biofilm. This is because sand filters, softeners and carbon filters all present extensive amounts of rough surface area over which the water must flow. Primary control of biofilm formation on pre-treatment filtration processes is usually by recirculation and other engineering means which ensure constant movement of the water over the filter media. Secondary control is by frequent back-washing — re-directing the output water back through the filter in the opposite direction to the normal flow. The media become disturbed so that the biofilm and other particles break off into the water flow and are dumped to drain. Sand and carbon filters, along with softeners need their output to be dumped to drain for a period after back-washing. Modern pre-filtration systems are designed and engineered so that these controls operate automatically either through electro-mechanical or computerised timers, valves, drains, etc.

Irrespective of the quality of the engineering controls it is not unusual to be able to trace microbiological contamination of pharmacopoeial waters to the pre-treatment systems. This is particularly true for carbon filters (Morin et al, 1996). Periodic high temperature sanitisation is commonly used to control biofilm formation in carbon filters. According to the construction material of their housings this may be done by back-washing with hot water or steam, or (rarely) steam under pressure. If microbiological problems persist there may be no practical alternative to replacing the carbon filter completely.

Preparation Processes for Purified Water

Numerous methods are available for preparing water to comply with the compendial limits applying to *Purified Water*. Modern technologies easily enable compliance with the chemical and physical limits. None of the methods available for preparing *Purified Water* have any significant effect on reducing the numbers of microbiological contaminants present in the feed-water. Some methods, if not adequately controlled, may worsen the microbiological quality of their output water.

- *Cation–anion exchange columns* — the most commonly used method of preparing *Purified Water*. The feed-water is purified by passage over ion-exchange resins. This method is effective, robust and time-proven.

  It does however present an extensive surface area on which biofilm may form. This is controlled principally by two means. First, practical pharmacopoeial cation–anion exchange systems should be engineered to ensure that water is in constant motion over them. This is usually achieved by re-circulating a portion of their output water. Second, and perhaps more importantly, cation–anion exchange resins become 'exhausted' over time and require regeneration with concentrated acids and alkalis. Microorganisms forming as biofilms in the columns are killed in the process of regeneration.

  Constant movement and frequent regeneration are necessary to maintain a high quality microbial output from cation–anion exchange columns. For microbiological purposes, it is generally insufficient to re-generate the columns only when they show signs of them becoming exhausted (ie from seeing an increase in the conductivity of the output water), therefore it is customary practice for regeneration to be carried out at time intervals, eg at 48 or 72 hour time intervals, according to experience and what has been proven through validation. In some plants, twin units may be installed to ensure that capacity demands do not over-extend the interval between regenerations.

- *Mixed-bed ion exchange units* — these apply the same principles as cation–anion exchange columns, with both types of resin contained in one unit.

Microbiologically they have the same potential problem of biofilm formation on the resins as cation–anion columns and share the same solutions.

- *Electro-deionisation (EDI) (continuous deionisation)* — these water preparation processes, or continuous deionisation (CDI) as they are sometimes called, do not require regeneration with strong acids and alkalis. In industry this is perceived as advantageous from the Health and Safety standpoint.

  A stack of plates across which an electrical current is applied to give different polarities to adjacent plates is at the base of EDI units. The 'compartments' between alternate sets of oppositely charged plates contain ion-exchange membranes.

  Deionisation is by two means — the first by ion exchange within the membranes, and the second by which ions in solution are attracted to electrodes with the opposite electrical charge. In EDI, the ion-exchange membranes do not require acid and alkali regeneration because H^+ and OH^- ions which are continuously produced from water by the electrical current serve this function.

  EDI was first commercialised in the late 1980s and although this may not seem overly recent, the pharmaceutical industry is very conservative and most applications of EDI to date have been as 'polishing' units in association with RO (see below).

  EDI units are not known to have particular issues in relation to microbiology. Sanitisation may be done in line with upstream RO units using the same chemical processes indicated below. Alternatively they can be sanitised with hot water, but this may limit the life of the EDI plate stacks (at best a three-yearly replacement frequency should be considered) due to deleterious effects on the membranes. It may take anything from six to 36 hours after sanitisation for the water quality output from EDI units to equilibrate to the conductivity and total organic carbon (TOC) levels seen before sanitisation.

- *Reverse osmosis* — works like a molecular sieve in which microorganisms and endotoxins are too large to pass through semi-permeable membranes (Klumb, 1975).

  Semi-permeable membranes only allow small molecules, such as water molecules, to move through them. When two aqueous solutions with different ion concentrations are separated by a semi-permeable membrane, the effect of osmotic pressure is such that water passes through the membrane in the direction of the more concentrated solution.

In RO a pressure sufficient to overcome the osmotic pressure is applied to the compartment with the higher ionic concentration such that the net movement of water molecules is now from the area of high solute concentration to that of low solute concentration.

Most semi-permeable membranes used in commercial RO units are made from cellulose acetate, polysulfonate, or polyamide, spirally wound around hollow tubes and assembled into pressure housings. Twinned units (twin-pass) are frequently used in pharmaceutical applications, with part of the output from the second unit diluting the feed-water to the first unit.

It has been reported that bacteria can "grow" through semi-permeable membranes, but this has never been proven. RO membranes are subject to 'fouling' — ions and biofilm may accumulate on the upstream side. It is possible that microorganisms from these biofilms may penetrate to the downstream side of RO units via the mechanical seals, O-rings, etc that are inevitable in practical applications. To counteract this, a flush cycle applied to the upstream side is often used to reduce build-up of scale and biofilm.

RO systems have to be chemically sanitised to remove biofilm once formed. Prior to sanitization, it is important to clean the first-pass RO system using a two-stage chemical treatment involving an acid cleaner, such as citric acid, followed by sodium hydroxide. Sanitization is usually done with formaldehyde, hydrogen peroxide, or peracetic acid. With two-pass RO systems, the second pass is not likely to require cleaning as frequently as the first.

Preparation Processes for Water for Injection(s)

Compliance with the limit applying to endotoxins significantly restricts the methods available for preparing *Water for Injection(s)* and *Highly Purified Water PhEur*. Microorganisms do not pass over distillation columns or through RO membranes, nor do endotoxins (Juberg, 1977).

PhEur only allows *Water for Injections* to be prepared using distillation. This is somewhat of an anomaly among pharmacopoeial monographs which normally contain only a specification against which samples of the material described can be tested for compliance. *USP* is more liberal in that it allows *Water for Injection* to be prepared using distillation or RO.

In a move towards harmonisation with *USP*, *PhEur* has introduced the monograph *Highly Purified Water* which has exactly the same specification as *Water for Injection(s)*, but has no restrictions on the means of preparation.

- *Distillation* — the temperatures achieved in distillation kill most microorganisms, and any which are not killed are, like endotoxins, too heavy to be carried over the still with the water vapour. From a microbiological standpoint, distillation is a virtually foolproof method of preparation of pharmacopoeial grade waters.

  Stills may be limited in their ability to prevent endotoxin carry-over. For this reason, it may be advisable to ensure that the endotoxin challenge to stills is not excessively high.

  It has become customary in the pharmaceutical industry — but not a regulatory or compendial requirement — for the feed-water to stills used for preparing *Water for Injection(s)* to comply with the requirements for *Purified Water*, thus minimising the microbiological and potentially the endotoxin challenge. The other factor potentially influencing the endotoxin challenge to stills is 'blow down'. This is the name given to the mechanism whereby the residual ionic and organic matter which is not carried over the still with the water vapour is removed either continuously or intermittently to prevent primarily build-up of scale within the still.

- *Reverse osmosis* — has been described above as a method of preparation of *Purified Water*. It may also be used to prepare *Water for Injection USP* and *Highly Purified Water PhEur*, but not for preparation of *Water for Injections PhEur*.

  RO is capable of producing output of the required endotoxin standards, and may be less energy intensive and therefore more economical to run than distillation. Capital costs may be comparable. Validation and monitoring costs may be higher.

  RO can be a very attractive proposition for active pharmaceutical ingredient manufacturing processes where very large volumes of water are required and where there is no explicit requirement to use *Water for Injection(s)* but at the same time there is a requirement to provide assurance that the end-product does not carry a significant endotoxin burden. In manufacture of parenteral dosage forms, RO water meeting the specifications for *Water for Injection(s)* may be useful for washing and clean-in-place systems for vessels and equipment.

Storage and Distribution

Storage and distribution systems are the means by which pharmacopoeial grade water is provided to the user, where, when, and in the quantities required. The central problem with storage and distribution systems is the potential for water prepared to comply with satisfactory microbiological quality standards being compromised en route to the user. The tanks, pipe-work, pumps, valves and other ancillary equipment required for storage and distribution present extensive surface areas on which biofilms can form.

The proper design and construction of storage and distribution systems for constant flow and for smooth internal surface finishes is therefore central to the successful supply of pharmacopoeial grade waters of satisfactory microbiological quality.

Before addressing the main components of storage and distribution systems it is important to consider the materials of construction.

The first microbiological consideration which must be given to the choice of materials is internal smoothness. Most pharmacopoeial grade water systems are made from 316L stainless steel, but some are constructed with plastic pipe-work. A recommended starting point for internal finishes for stainless steel water pipe-work is not more than 0.5 μm Ra. Plastic pipe-work (eg UPVC) tends to yield slightly higher roughness average values than stainless steel, but this may not be a true index for comparing different materials. There are, for instance, data available indicating that biofilm attaches more poorly to plastics than to stainless steel.

Standard ranges of stainless steel tubing are available, defined by the manner of finishing the internal surfaces ('cold drawn', 'honed', 'polished', 'electro-polished'). Pharmacopoeial grade water systems operate satisfactorily with all these types of pipe-work although 'cold drawn' tubing, some even with certified internal finishes of less than 0.5 μm Ra, would now only be thought appropriate in older installations.

Internal smoothness is not purely a function of material quality, but also of how materials fit together to make up complex storage and distribution systems. It is important that roughened internal areas are not created when lengths of tubing of the same material are joined together, or when tubing made of one type of material has to be joined to tanks, pumps, valves, etc of a different type of material. For instance, the Ra value can be used to specify the internal surface finishes of both stainless steel tubing and storage tanks for maximum compatibility. With systems using mixed components it may be more difficult to create compatible specifications.

- *Storage tanks*
  Storage tanks are key parts of water systems, existing to ensure there is enough water available to satisfy demand (particularly peak demand) without

the system emptying. However, for microbiological reasons the 'mean residence time' for water in the tanks should be minimised.

In the best designed systems the storage tank capacity is based on the requirements of an approximate demand matrix from each off-take point throughout the day, and over the days of the working week. Over-capacity introduces the risk of water remaining stagnant within tanks for unacceptably long periods of time. Tanks are usually allowed to fill to a particular level (the 'fill level'), and then for the water from the preparation process to be recycled back through preparation or to be dumped to drain. For new projects, engineers may deliberately design for spare tank capacity to allow future flexibility and growth; in such cases the fill level should be set appropriately for the demand. If demand lessens, fill levels should be reduced; if demand increases fill levels may be increased.

There is always an area of tank surface which is not submerged beneath the water level. This differs from a minimum area when the tank is full and no more freshly prepared water is being added, to a maximum area when draw-off has emptied the tank to its over-capacity safety margin. To prevent biofilm forming on these wet surfaces, the water from the distribution pipe-work is recirculated into the tank via a spray ball or balls. The returning water forced out from the spray balls hits the upper internal surfaces of the tank with so much force that biofilm cannot form. The force of the returning water also ensures a constant movement of water over the other periodically exposed internal surfaces of the tank. Water is never stagnant in properly designed tanks.

There has been much regulatory concern (Anon, 1993) over the venting of tanks. Clearly, storage tanks in which water levels rise and fall must be vented to avoid the risks of buckling inwards or splitting. The regulatory requirement is for these vents to be protected by bacteria-retentive filters.

The application of bacteria-retentive filters to tank vents introduces problems, both technical and logical. The technical problems are two-fold — which orientation to install the filters, and how to prevent condensation of water on the filters leading to secondary biofilm problems.

Hydrophobic filters are designed to operate in one direction only with gas flow from the non-sterile to the sterile side. In venting, flow is in both directions. Vent filters are properly installed such that the direction of flow is from the outside to the inside of tanks.

If water condenses on filters this may lead to blockage of the filter and/or to biofilm formation. It is best to install vent filters in heated housings to prevent these possibilities.

The logical problem is in the use of bacteria-retentive filters on non-sterile systems. Of course it makes sense that water storage and distribution systems should not be open to contamination from the general environment, but does it make sense to take on all the other regulatory obligations ordinarily associated with bacteria-retentive filters and protection of sterility? It has become common practice to test these filters for integrity before and after installation — again there is no point installing a filter if there is no assurance that it has been fitted properly into its housing and its media is undamaged. Problems only arise if these integrity tests fail (and end-of-use testing may be after three or six months of use, or even longer) — what action should be taken on the product which has been manufactured with water from an unprotected system, what action should be taken on the product-contact manufacturing equipment cleaned with water taken from an unprotected system? The answer, of course, is no that no action on product is required but the outcome of the documented investigation may call for introduction of some improvements to the type of filter or housing used, the process of installation, the protection given to the vent filter in use, etc.

- *Pipe-work*

  The reality of pipe-work can easily differ from its appearance on engineering drawings. This is because water system design typically takes little account of the practicalities of structural pillars, ceiling clearances, corridors, available space, personnel access, etc in actual manufacturing facilities. These problems are addressed during building, either by the sub-contractors alone or in conjunction with local site engineering.

  FDA (Anon, 1993) expects the engineering drawings of water systems to be compared with actuality on an annual basis.

  The consequences of often *ad hoc* practical solutions to real construction problems may be sub-optimal microbiological design. Pipe-work may not have been constructed to slope to drain, valves may be mounted in the wrong orientation, and there may be branches from the pipe-work where there is no water movement or less water movement than intended — these regions are called 'dead legs'.

  Branch pipes are usually required at user points, or around pumps, etc. When water is allowed to stand in these branch pipes for protracted periods of time the opportunities for contamination by adventitious microorganisms increases, along with with the potential for bioburden formation. The maximum allowable branch pipe dead leg length based on the relationship between the diameter of the main pipe through which the water flows and the diameter of branch pipe has been defined (FDA, 1993). This definition should not be interpreted too literally — a short dead leg on a rarely used user point may well be more prone to bioburden formation than a longer dead leg through which water draw-off is very frequent. The only sound advice is to minimise all dead legs.

Bioburden arising from poor quality welding may also form on rough internal surfaces of pipe-work. Automated inert gas blanketed welding techniques minimise this possibility.

- *Pumps, valves, ancillary equipment*
  The practicalities of water distribution systems demand that water is pumped around the system, that it can be drawn off where and when required, and that quality characteristics can be monitored (often continuously). All this necessitates some compromise of the principle of water flowing rapidly and continuously over smooth and uninterrupted internal surfaces.

  Pumps are required to have a sanitary centrifugal design in which the drive motor never comes into contact with the water. Generally these are microbiologically reliable although sometimes biofilm has been known to form at seals, etc.

  Microbiological problems arise generally where more than one pump is used. From a business risk standpoint, sanitary pumps on water systems are highly critical — no pump means no water, and no water means no manufacture. This criticality can be addressed by one or a combination of several means.

  — A stand-by pump completely independent of the water system and its plumbing can be provided. On breakdown of the primary pump, it can be quickly plumbed in. However, if the stand-by pump is stored in the damp environment of a water plant room it will become biofilm contaminated. This can be avoided by filling the stand-by pump with an alcoholic disinfectant or storing it elsewhere in a dry environment.

  — Twin pumps can be installed in the water system, with either one operational and the other stand-by, or with both alternating. From a biofilm protection standpoint it is best for both pumps to have water flowing continuously through them — this can be done by having a drain line through which water flows when a pump is non-operational.

The plumbing between two such pipes should be installed with care to avoid dead legs.

Valves should be of sanitary design. Sanitary diaphragm valves are sometimes referred to as 'zero dead leg' valves, but 'zero' is a misnomer as there is always the potential for some dead leg and standing water.

Sanitary valves contain a flexible diaphragm. Water flow is stopped when the diaphragm is screwed down onto a 'weir' located on the opposite side of the pipe to the screw bonnet. When the diaphragm is released from the weir water flows through the valve. These valves can be operated manually or can be motorised for automated operation.

While acknowledging that diaphragm valves are better than any alternative available, two main types of microbiological problem can arise from their misuse:

— Mis-orientation: if diaphragm valves are mounted bonnet-upwards on horizontal stretches of pipe-work stagnant water (or severely restricted water flow) may present behind the weir.

— Damage: if the plasticity of the diaphragm is lost (possibly due to aging or exposure to high temperatures in high temperature re-circulating water systems, or to high temperature sanitisation) it may crack. This may not be obvious from leakage as the diaphragm is generally supported by a rubber seal. The possibility arises of biofilm forming between the diaphragm and its seal. This may contaminate water drawn off through the valve, and contaminate the water system itself by sloughing of microorganisms into the pipe-work. A replacement strategy is recommended within the preventive maintenance programme.

All other accessories within water systems present risks of biofilm formation, and only those which are absolutely necessary should be installed. Clearly there must be level devices within tanks, temperature, conductivity and total organic carbon sensors and samplers within pipe-work.

FDA (Anon, 1993) has advised strongly against the use of bacteria-retentive filters within water systems due to their potential to block and restrict water flow as biofilm forms on the extended surfaces of their pleats.

Figure 1.2 Generalised Re-circulating Purified Water System

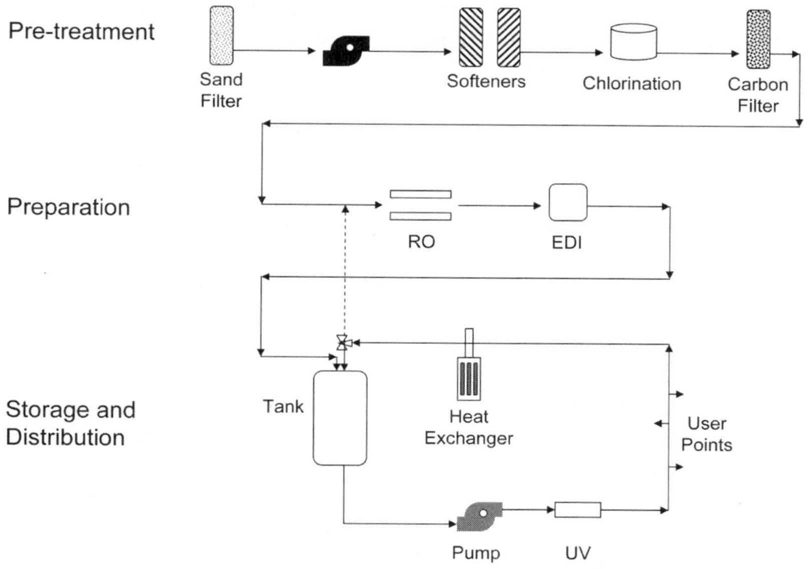

Water System Design

The essential elements of pharmacopoeial water systems have been portrayed thus far as if they were jig-saw puzzle pieces. The manner in which they are put together is of course integral to microbiological control (Collentro, 2002) .

Figure 1.2 represents a very simple re-circulating *Purified Water* system following the principles of pre-treatment, preparation and finally storage and distribution. Most pharmacopoeial water systems are re-circulating systems.

In this model, water is pumped from the storage tank, round the distribution loop, and back to the tank. In some circumstances subsidiary loops and additional pumps may be included in the design. It should be recognised that water movement must be maintained in the whole system at all times, with the limits on flow rates (usually 1–3 m per sec) applying at the point on the distribution loop where water is expected to be slowest at times when draw off is maximum.

A heat exchanger is depicted close to the return to the tank. This may serve two purposes.

- In ambient temperature distribution systems, the temperature of the water may increase over protracted periods in which there is no water draw off, eg over shut-downs, long weekends, etc. To control this, some 'trim-cooling' through a heat exchanger, linked to a temperature sensor which triggers its operation when the temperature exceeds a pre-set level, may be necessary.

  In high temperature re-circulating systems such as those required for *Water for Injection(s)* a heat exchanger near the 'return' of the distribution loop serves to maintain the temperature above 75°C or 80°C, or whatever lower limit has been set on water temperature throughout the system. High temperature re-circulation is only rarely seen in *Purified Water* systems where microbiological limits should easily be achievable without recourse to additional energy expense.

  High temperature re-circulating systems must have cooling at the user points, as temperatures above 75°C are too high to be practical for most purposes (eg compounding of formulations is generally done at 20°C). In some other circumstances, eg rinsing machine parts after cleaning, they may actually be hazardous to personnel. User point cooling by means of localised heat exchangers is essential for *Water for Injection(s)*.

- The heat exchanger may also be used to raise the temperature of the water to sanitise the system — this is the most robust method of water system sanitisation. Temperatures over 65°C for relatively short exposure times kill gram-negative microorganisms and no chemical residues are left behind.

  High temperature sanitisation is best suited to stainless steel distribution systems. Plastic pipe-work may not withstand high temperatures. As it may take some hours to raise the temperature in a large water system, sanitisation is usually done at times when there is no draw off, eg overnight or at weekends. Temperatures of over 80°C are usually specified with dwell times dictated by the time taken for the water to heat up and cool down, and the time available for sanitisation of the system.

  It should be recognised that *Purified Water* systems are not sterile. In fact they are being constantly inoculated with fresh microorganisms from upstream. The power of sanitisation (in this case the temperature achieved and maintained, whether 80°C for one hour or 100°C for 30 minutes) is less important to ensuring a good microbiological quality than the frequency of sanitisation.

Although heat is the most commonly used means of sanitising water systems it is not the only industrial method. For instance, it may be completely unsuited to some types of plastic pipe-work. For this reason, other means of sanitisation have been developed. Those which do not leave chemical residues, eg ozone (Burkhart et al, 1996), which break down to water and oxygen, as does hydrogen peroxide, are preferred. Peracetic acid has been also been known to be used.

Many re-circulating systems include UV lights within the loop. It is probably fair to say that in pharmacopoeial water systems, UV irradiation is a solution in search of a problem; but has nonetheless become commonplace. In properly designed and constructed water systems good microbiological control should be achievable through provision of smooth, dead-leg free internal surfaces, high flow rates and periodic sanitisation without the need for UV light.

In fact, UV lights may disguise rather than fix problems. It is scientifically well known that UV light damages DNA (Harm, 1980). It creates covalent linkages between bases that prevent replication and transcription, causing death when not repaired by the cell. However, many bacteria have developed repair mechanisms to compensate for the damaging effects of UV radiation (Ganesan and Smith, 1968; Zimmer and Slawson, 2002; Piluso and Moffatt-Smith, 2006). In other words, the effect of UV light may be to reduce the numbers of microorganisms recoverable from water systems by conventional culturing techniques, when in fact higher numbers may be undergoing repair within the system.

Although re-circulating water systems are most common in pharmaceutical applications, straight-through systems are not prohibited. These may range from quite small for occasional batching, to massive systems used by large volume parenteral manufacturers who use as much water as can be supplied.

A small system is depicted in Figure 1.3. The means of preparation is RO with an EDI polisher. An automated valve linked to a conductivity circuit dumps the output of the RO system to drain until it complies with limits, before it is allowed to be used as in-feed to the electro-deioniser. A second automated valve dumps the output of the electro-deioniser until it also complies. By this means the systems itself ensures that any water released for use complies with the pharmacopoeial limits on conductivity. There is no automated microbiological assurance, and the avoidance of biofilm formation in such systems relies on a constant bleed to drain through both processes.

Figure 1.3 'Straight-through' Purified Water System with Conductivity Controlled Valves

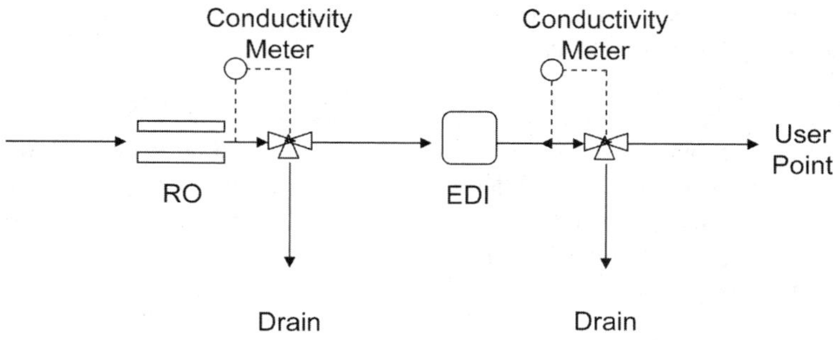

Microbiological Monitoring of Water Systems

The principle decisions to be made with respect to microbiological monitoring of water systems are where, when, and how often. The best monitoring programmes are the ones where there is good reason for having chosen where, when and how often to take samples. The comments here apply to monitoring downstream of the preparation process(es).

FDA (Anon, 1993) expects pharmacopoeial grade water samples to be taken via user points (meaning all user points) because this gives an indication of the quality of the water actually used. Although this would seem quite straightforward, it has its complications.

- In some hard-plumbed situations, for instance pipe-work hard-plumber via an automatic valve to a manufacturing vessel, there may be no access to the user point. The best solution here is to install a valve which incorporates a sample off-take, such that the water sample has passed through the valve in order to detect any contamination of the valve itself, as well any contamination of the water system.

- As already indicated, user point valves may themselves become contaminated. As a consequence to sample only, through user points may give an impression of the quality of water being used, but no indication if contamination exists throughout the water system, or only in the user point sampled. For this reason, samples are usually not only taken at user points, but through specifically included system sample points.

The location of system sample points should be considered in relation to understanding where things are mostly likely to have 'gone wrong' when they go wrong (as the inevitably do at least sometimes). For this reason, system sample points are at the very least placed downstream of the pre-treatment and preparation processes, and at the 'return' of the water distribution loop to the storage tank.

Monitoring is a type of 'insurance policy', telling you that something has gone wrong, and hopefully where this has most likely occurred. At extra cost, system sample points beyond the minimum can in many facilities be seen downstream of each stage of pre-treatment, and at intermediate stages in distribution loops. It makes sense to monitor downstream of the pump in a distribution loop.

It makes little sense, although it is frequently seen, to monitor downstream of a UV light, except to guarantee very good data. Microbiological technique is too insensitive to detect marginal inadequacies in sterilisation processes — this is well known in the context of the pharmacopoeial *Test for Sterility* and the same limitations apply even more so to the idea of testing a small sample — 1mL, 10mL, 200 mL — of water by a bioburden technique downstream of a UV light.

How often should monitoring samples be taken?

When water, either *Purified Water* or *Water for Injection(s)* is used for compounding pharmaceutical formulations, it makes sense from a compliance standpoint to have microbiological monitoring data available from each date on which formulation occurs.

Endotoxin data, unlike microbiological data, can be obtained within a few hours of sampling. When endotoxin is tested from *Water for Injection(s)* compounding user points, it makes sense to sample as early in the day as possible, to allow re-sampling and testing in 'real time' when necessary.

Beyond this, it is difficult to generalise (Cooper & Polk, 1998) on how often each sample point and user point should be monitored. This is a function of individual systems, their design, history of operation, history of monitoring, and the ways in which they are used. In every facility there will be subtleties but the following factors should always be considered.

- The sanitisation frequency: assuming sanitisation is at least partially effective in removing microorganisms from pharmacopoeial water systems, the lowest probability of detecting viable microorganisms should occur immediately after sanitisation and the highest probability immediately before sanitisation. In *Water for Injection(s)* systems it is quite possible that the probability of detecting endotoxin is the other way round, ie greatest immediately after sanitisation and lowest immediately before sanitisation.

- 'And on the seventh day God ended his work which he had made; and he rested on the seventh day from all his work which he had made' (Genesis Chapter 2 verse 1). Although some pharmaceutical manufacturing facilities work 24 hours a day, 7 days a week, it is fair to say that most do not. In the working week, water systems operate in a 'steady state' where the water taken off is replaced with freshly prepared water. At weekends, there may be far more re-circulation of water, and at the beginning of the working week equipment may change from a stand-by mode to a fully operational one. These factors can have an effect on microbiological quality, so it is probably reasonable to sample all system points at least on a weekly basis and preferably at the beginning of the week immediately after the weekend break.

- Undefined manufacturing-related variables potentially affecting user points should influence their sampling frequency. Over a defined period, say three or six months, there should be some data applying to each user point from each day of the week. For instance, if there are six working days each week, weekly sampling would allow sampling on each day twice in three months, or fortnightly sampling would allow sampling on each day twice in six months.

Water samples should be tested promptly after collection. Most companies include some allowance in their procedures along the lines of 'if the sample cannot be tested within *x* hours of collection it may be held in a refrigerator at 2–8°C for up to *y* hours before testing'. Recourse to this provision should be infrequent and where it is applied, should be noted on the record sheets. The provision is necessary for practical reasons, probably originating in *Standard Methods for the Examination of Water and Wastewater*, where, in the 20th edition (Anon, 1998) it states that if samples cannot be tested within one hour they should be 'iced' for transportation, and there should be not more than 30 hours between sampling and testing.

Microorganisms recovered from water testing should be identified. In *Water for Injection(s)* this should not lead to an onerous workload as the typical recovery is nil per 200 mL sample tested.

Some microorganisms are, however, generally recovered from many, but not necessary all, samples of *Purified Water*. Technicians should at least be trained and assessed in their abilities to distinguish gram-positive from gram-negative types on the basis of colonial appearances on agar media. It is generally recommended that representative colonies from each sample are identified to species level, for which a variety of automated and semi-automated techniques are available. The majority of types recovered should be gram-negative with perhaps a few gram-positive spore-formers. Gram-positive bacteria of human origin should throw suspicion on sampling and testing technique.

Many of the available identification techniques, particularly those operating at nucleic acid levels, give very precise identities. The value of these detailed identities is questionable. Certainly the data should be analysed to obtain a picture of the typical microflora. It should not be surprising if this includes pathogenic species such as *Pseudomonas aeruginosa* and *Burkholderia cepacia*, but they should certainly not be among the most commonly recovered species. It is very tempting to over-react to individual recovery of pathogenic gram-negative species, but serious reaction should be confined to reoccurring isolations. Intestinal gram-negative bacteria, such as the coliforms should not be recovered, as they are indicative of a poor level of hygiene which should never occur in pharmaceutical manufacturing facilities.

Compendial standards for microorganisms in water are based on conventional media incubation methods. However in recent years, there have been major developments towards developing faster microbiological techniques (Green and Randell, 2004). Direct epifluorescence technologies (DEFT) have been found particularly suitable and are considered potentially valuable for testing pharmacopoeial grade waters.

The critical question is — why ?

At a cost of up to Euro 80,000 DEFT apparatus is a mere drop in the ocean to pharmaceutical companies earning Euro 8 billion pre-tax profit per annum, but it is also the price of a pretty decent Mercedes. It is often assumed that there is something intrinsically valuable in having quick rather than slow results, but in water testing this is just not true.

Water systems are neither controlled nor controllable through microbiological testing. They are controlled through proper design, construction and operation. Like a super-tanker they take a long time to start, stop or change direction. In *Water for Injection(s)* systems, out-of-specification results are almost totally unheard of. The author would be very surprised if out of specification results rose above the 1% mark in *Purified Water* systems. With such stability it really does not matter if monitoring results are obtainable in four hours or four days, they are still odds-on to be satisfactory. Add to this the fact that DEFT only provides data against numerical limits such that any knowledge of the types of microorganisms present still requires conventional media incubation techniques, and it should be apparent that quicker results may not provide enough benefits to merit the expenditure.

The author acknowledges that these views may not be acceptable to all who have invested much time and energy in developing these techniques, and that his opinion is based on currently available technologies and not on what might be available in the future. In support of rapid methods, however, they may well be useful is as investigational tools to assist in correcting water systems which have become seriously microbiologically unstable. As already mentioned, water systems are as difficult to manoeuvre as super-tankers, especially when the direction in which they are heading becomes apparent only four or more days after an adjustment has been made. Availability of data within a few hours may in these circumstances be of significant benefit.

Investigating Excursions versus Microbiological Limits for Water

All excursions beyond microbiological limits for waters should be investigated. It is possible, but probably unlikely, that a root cause may be found.

The most obvious cause of excursions in water monitoring is contamination in sampling or testing. When the types of microorganisms are more likely associated with personnel (gram-positive cocci) than with water (gram-negative rods), the root cause most likely lies in sampling or testing.

The cause of genuine excursions must lie in the facility. This is where the investigation should focus, although some laboratory testing may be valuable. It should be accepted that actually finding a root cause for a microbiological excursion is often very difficult and the most likely outcome of investigations is 'inconclusive'.

The value of investigation is that many previously unknown faults may be noticed and corrected. In fact there is a higher probability of noticing and correcting leaks, hoses which have become permanently attached through custom and practice, hoses dangling in sinks, drains which have become blocked, etc, as result of investigation than as a result of regulatory inspection. That cannot be a bad thing.

OTHER SOURCES OF MICROBIAL CONTAMINATION FROM WATER

Pharmacopoeial grade water may not be the only potential source of contamination by water-borne microorganisms in pharmaceutical manufacturing facilities. These other sources are less predictable and often more difficult to control.

Although mandatory to compound dosage forms and the final stages of equipment cleaning with waters of pharmacopoeial grade, waters of other qualities are regularly used in manufacturing facilities.

- When equipment presents cleaning difficulties (eg ointment manufacturing equipment) and/or where the scale of manufacture is large (eg in oral liquids or API manufacture) etc, it is customary to use potable water in the first stages of cleaning. This water is required to meet public health standards applying to drinking and dietetic purposes and may in practice be of better or worse microbiological quality than *Purified Water* (particularly if it has been chlorinated and is used hot).

- Hand-washing is mandatory for personnel working in pharmaceutical facilities. Where prevention of microbiological contamination is critical (eg in the aseptic manufacture of sterile products) it is often better to use non-aqueous hand washing techniques close to protected areas. The detailed regulatory requirements applying to the transfer of microorganisms in air (HEPA filtration, air flow direction, pressure differentials, self-closing doors, etc) have no bearing on the potential transfer of water-borne microorganisms by personnel from hand-wash stations to equipment and possibly to exposed products.

These other sources of water are more than likely to be directed to the same floor drains as pharmacopoeial grade waters. Drains are a secondary but very potent potential source of water-borne microorganisms. It must be accepted that drains will always be contaminated: it is only the extent to which they are contaminated and the protection afforded to the manufacturing operations around them that can be influenced.

Drains can be kept clean and fresh-smelling by periodically flooding them with sodium hypochlorite solution. Engineers may argue that mild steel drain pipes are corroded by sodium hypochlorite, but this is arguably the lesser of two evils — no pharmaceutical product has yet been recalled because of corroded drains in the manufacturing facility.

Figure 1.4 Generalised Arrangement of a Floor Drain

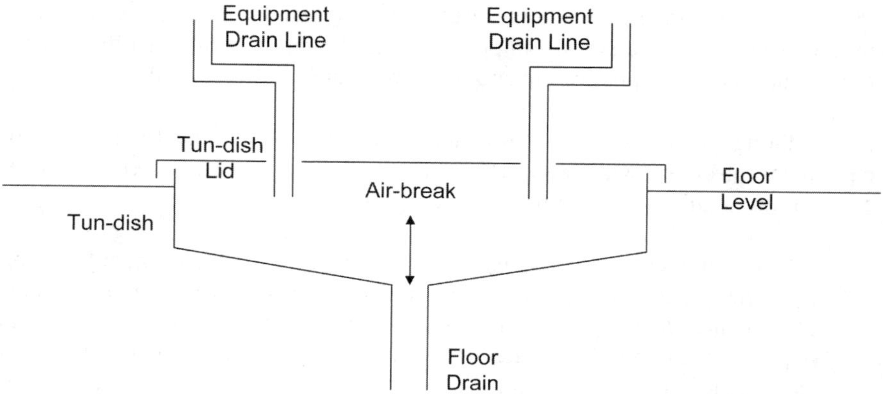

One way of protecting the manufacturing environment from contamination coming from the drains is by creating an air break between equipment drains and floor drains. Microorganisms only move in air as a result of air movement itself. Something like a 2 or 3 cm air break between the outlet of the equipment drain and the top of the floor drain is thought optimal. Water falling through longer distances may have a greater probability of splashing back and forming aerosols which could, at least in theory, re-contaminate the equipment and the surrounding environment.

Floor drains should be able to take water away at maximum challenges, for instance, at the end of a clean-in-place cycle where the total contents of a significantly large tank are discharged through a fully open drain valve. Some 'buffer' capacity is usually provided by so called tun-dishes (Figure 1.4) designed to hold the maximum challenge without water spreading over the floor around the drain.

Microorganisms cannot move in air, nor can they move on dry surfaces. Some can move (although probably not very far or very fast) on moist surfaces, and they will travel wherever water carries them. To prevent floors being wetted drains should be:

- equipped with tun-dishes of adequate dimensions and capacity

- located at the low point in floors — although this would seem obvious, it does not always happen

- visible — it may seem like good hygienic practice to conceal equipment, sink and floor drains behind smooth cleanable polished stainless steel panels — but this only prevents standing water due to drain blockage or under-capacity, or drains with inadequate tun-dishes being noticed. Problems which go unnoticed are still problems.

The final source of water-borne microorganisms in pharmaceutical manufacturing facilities is from condensate. Where steam is used in a process or is exhausted from an item of equipment condensate is likely to form on cooler surfaces. Where there is sufficient air movement to keep these areas cool steam will usually be removed before it condenses, but this may not always be the case, and certainly merits attention as a means of preventing contamination by water-borne microorganisms.

REFERENCES

Anon (1969) Department of Health and Social Security, Welsh Office, Department of Environment Reports on Public Health and Medical Subjects No. 71 *The Bacteriological Examination of Water Supplies*. London: HMSO.

Anon (1993) *FDA Guide to Inspections of High Purity Water Systems* Washington DC, USA: ORA, FDA.

Anon (1998) *Standard Methods for the Examination of Water and Wastewater* 20th Edition. American Water Works Association, Denver Colorado.

Burkhart, M., Wermelinger, J., Setz, W. and Muller, D. (1996) Suitability of polyvinylidene fluoride (PVDF) piping in pharmaceutical ultrapure water applications. *PDA Journal of Pharmaceutical Science and Technology* **50** (4) 246–251.

Collentro, A.W. (2002) Practical microbial control techniques for pharmaceutical water purification systems. *Ultra Pure Water* **19** 53–58.

Cooper, J.F. and Polk, C.S. (1998) Monitoring water systems for endotoxin. *CRL-LAL Times* **5** (2) 1–6.

Donlan, R.M. (2002) Biofims: microbial life on surfaces *Emerging and Infectious Diseases* **8** (9).

Dunne, W.M. (2002) Bacterial adhesion: seen any good biofilms lately? *Clinical Microbiology Reviews* **15** (2) 155–166.

Ganesan, A. K. and Smith, K.C. (1968). Dark recovery processes in *Escherichia coli* irradiated with ultraviolet light. *Journal of Bacteriology* **96** 365–371.

Green, S. and Randell, C (2004) Rapid microbiological methods explained. In Microbiological Contamination Control in *Pharmaceutical Clean Rooms* ed N. A. Halls CRC Press; Boca Raton, Florida.

Harm, W. (1980) Repair related phenomena. In *Biological Effects of Ultraviolet Radiation* ed. Harm, W. pp. 124–134. Cambridge: Cambridge University Press.

Hunt, S.M., Wemer, E.M., Huang, B., Hamilton, M.A and Stewart, P.S. (2004) Hypothesis for the role of nutrient starvation in biofilm detachment. *Applied and Environmental Microbiology* **70** (12) 7418–7425.

Juberg, D.L. (1977) Application of Reverse Osmosis for the Generation of Water for Injection *Bulletin of the Parenteral Drug Association* **31** 70–78.

Klumb, G. H. (1975) Reverse Osmosis — A Process in the Purification of Water for Parenteral Administration *Bulletin of the Parenteral Drug Association* **29** (5)

Morin, P., Camper, A., Jones, W., Gatel, D. and Goldman, J.C. (1996) Colonization and disinfection of biofilms hosting coliform-colonized carbon fines. *Applied and Environmental Microbiology* **62** (12) 44284432.

Piluso, L.G. and Moffatt-Smith, C. (2006) Disinfection using ultra-violet radiation as an antimicrobial agent: a review and synthesis of mechanisms and concerns. *PDA Journal of Pharmaceutical Science and Technology* **60** (1) 1–16.

Punakabutra, N., Numthapisud, P., Pisitkun, T., Tiranathanagul, K., Tungsanga, K. and Eiam-Ong, S. (2004) Comparison of different culture methods on bacterial recovery in hemodialysis fluids. *Journal of the Medical Association of Thailand* **87** (11) 1361–1367.

Reasoner D.J and Geldreich E.E. (1985) A new medium for the enumeration and subculture of bacteria from potable water. *Applied and Environmental Microbiology* **49** 1–7.

Stoodley, P., Wilson, S., Hall-Stoodley, L, Boyle, J.D., Lappin-Scott, H.M. and Costerton, J.W. (2001) Growth and detachment of cell clusters from mature mixed-species biofilms *Applied and Environmental Microbiology* **67** (12) 5608–5613.

Topping, L.E. (1937) The predominant microorganisms in soils (I) description and classification. *Zentralblatt fur Bakteriologie, Parasitenbunde, Infecktions-krankheiten und Hygiene (II) Abt*, Bd 97 No. 14/17, 287–304.

van der Linde, K., Lim, B.T., Rondeel, J.M.M., Antonissen, L.P.M.T. and de Jong, G.M.T. (1999) Improved bacteriological surveillance of haemodialysis fluids: a comparison between Tryptic soy agar and Reasoner's 2A media *Nephrol Dial Transplant* **14** 2433–2437.

Watnick, P. and Kolter, R. (2000) Biofilm, city of microbes. *Journal of Bacteriology* **182** (10) 2675–2679.

Zimmer, J.L. and Slawson, R.M. (2002) Potential repair of *Escherichia coli* DNA following exposure to UV radiation from both medium- and low-pressure UV sources used in drinking water treatment. *Applied and Environmental Microbiology* **68** (7) 3293–3299.

2

ISOLATION TECHNOLOGY

Tim Coles

1 INTRODUCTION

1.1 History and Development

The concept of placing pharmaceutical processes within enclosures, to protect either the product or the operator is not new. Gloveboxes of some sort have been used since modern pharmaceutical production began but this use declined with the advent of the high efficiency particulate air (HEPA) filter. This invention led to the construction of cleanrooms in which processes could be conveniently carried out with much reduced potential for contamination. In the late 1970s and the early 1980s however, the industrial climate started to change. The potency of active pharmaceutical ingredients (APIs) was increasing while there was a demand for improved operator safety and for higher product quality. Since the operators were at the heart of these issues, it was logical to move the processes into enclosures, which became known as 'isolators' in the pharmaceutical industry. This title distinguishes them to some extent from the 'gloveboxes' of the nuclear industry and the Class III safety cabinets of the biotech industry.

Over the years, various definitions have been offered for isolators but perhaps the most comprehensive description is to be found in *Pharmaceutical Isolators* (Midcalf et al, 2004). For the purposes of this chapter however, we may define the isolator as placing a physical barrier between the operator and the process. The object of this barrier may be to protect either the operator or the process and in some cases, both. Any air (or other gas) entering or leaving the isolator must do so through HEPA filters. Any material or product transfer process must maintain the barrier.

A vial filling line installed in 2004 in the UK. The isolator system was constructed by Skan to house a filling machine built by IMA, and is fitted with a dedicated VPHP gassing system.

The quality of the isolator environment should be defined in terms of the process requirements, such as particle burden, microbiological burden, humidity, etc. Thus the isolator can be used very much as a tool designed to achieve compliance.

This definition would though, exclude isolators fitted with devices such as 'mouseholes' (see Section 3, p57). These are basically minimum-sized holes in the isolator wall through which items such as filled vials can move continuously. We may perhaps stretch the definition to embrace devices which do theoretically allow some unfiltered air exchange, but under strictly controlled conditions. Beyond this we have the devices known as restricted access barrier systems (RABS) (see Section 7, p88).

Isolators have developed into various alternative forms with positive pressure, negative pressure, turbulent airflow, unidirectional airflow, flexible film construction, rigid construction, handling via gloves, handling via half-suits, standard design and custom design. A number of devices have been produced to transfer materials into isolators and to remove products, without breaking the containment. Most recently, methods have been developed to 'sterilise' isolators using gas, and in particular using vapour phase hydrogen peroxide (VPHP). All these issues are discussed further on in this chapter, though it may be noted here that the term 'sterilise' is not strictly accurate. This is because only heat (wet or dry) and irradiation produce a condition in which there is an absence of viable micro-organisms. The effect of chemicals is not absolute, and so is termed 'sanitisation'.

1.2 Typical Isolator Applications

1.2.1 Perhaps the most sophisticated of pharmaceutical applications is in aseptic filling. Here the entire filling line from the depyrogenation tunnel through to the capping machine may be enclosed in an isolator system. This will normally be built specifically to enclose the filling equipment, with transfer systems fitted to introduce stoppers and product, and gassing systems fitted for overall sanitisation, prior to processing. The line may also accommodate intermediate processes such as freeze-drying, where required. Isolators are used mainly to house the filling of vials but syringes and ampoules can also be accommodated. Isolated filling lines are particularly useful for cytotoxic products where the need for aseptic conditions must be combined with secure operator protection. Thus compliance can be achieved not only for product standards but also for health and safety standards.

It is worth noting at this stage that a U.S. Food and Drug Administration (FDA) inspector (Rick Freidman) has been asked on a number of recent occasions how any new filling line should be installed. His answer has been that he, and thus by implication the FDA, would expect such lines to be built with isolator technology (Lysfjord, 2004).

A flexible-film isolator system for sterility testing. The right-hand isolator is used for test work while the left-hand unit is used for VPHP gassing and transfer.

1.2.2 The next most frequent application for isolators in the pharmaceutical industry is probably in sterility-testing. The combination of closed testing methods such as the Millipore Steritest™ system with gassed isolators has virtually eliminated the occurrence of false positives (incidental contamination arising as a result of the test process itself). Relatively simple flexible film isolators can be used, with the isolator-dedicated version of the Steritest pump unit built into the base tray. Depending on the scale of the work, gloves or half-suits may be used for handling. An isolator linked with simple doors is often used to gas materials and tools before transfer into the main working chamber. In some cases, sterility testing isolators have been directly linked to autoclaves though this seems unnecessarily complex because the batch of samples has to be gassed into the isolator in any event.

1.2.3 Other applications in the pharmaceutical industry include either contained or aseptic loading and discharge of process equipment such as freeze-dryers, autoclaves, reactors, tablet presses, blender-dryers, and micronisers. Isolators are finding increased use in research facilities where materials of unknown toxicity may be handled.

1.2.4 Slightly aside from the pharmaceutical industry, isolators are now widely used in hospital pharmacies in the UK for total parenteral nutrition (TPN) compounding, IV additive preparation and cytotoxic reconstitution. They are also used for radio-pharmacy applications such as blood-labelling and increasingly in the preparation of positron emission tomography (PET) scanning solutions.

1.3 Pros and Cons

The main aim of an isolator is normally the improvement of product quality. By effectively removing one of the primary sources of contamination — the operators — the risk of contamination can be reduced by some orders of magnitude. Provided that the isolator system is operated correctly, this is certainly true and is the driving force behind many isolated aseptic filling lines.

Perhaps the next aim of isolator use is as a safety device. Clearly by separating a process handling toxic or pathogenic materials from the operators, very significant improvements may be made in operator safety. Again this is certainly true and is, for example, part of the driving force behind isolator systems for cytotoxic products.

Isolators have the capacity to provide specialised environments such as inert or low-humidity atmosphere although this attribute seems to be rarely used.

On the negative side, the installation of an isolator is likely to make any process less easy to operate and of course this is especially true of any process requiring much manual input. Working through windows using gloves and sleeves or half-suits is always going to be more difficult and slower than working on the open cleanroom bench or in an open-fronted laminar flow cabinet.

Then there is the issue of cost. At the dawn of isolation technology, it was envisaged that isolators would be installed in nothing more than laboratory-type environments and thus would save the great cost of cleanroom installation. In practice, the existing guidelines (Lee and Midcalf, 1994) quite reasonably propose that isolators are installed in some form of controlled environment. The lowest grade of controlled environment was then the GMP Grade D (at least in the UK) and so this became the requirement for isolator rooms. The pharmaceutical industry tends to the view that if a little of something is good, then more must be better. The result is that isolators are now frequently installed in ISO Class 7 or 5 cleanrooms.

Where negative pressure aseptic isolators are in use, for example, in radio- and hospital pharmacy cytotoxic units, this view may be justified. For normal positive pressure isolator operation, the use of full cleanrooms is perhaps less easy to justify. Furthermore, we then find operators wearing full cleanroom clothing, which further hampers their work in isolators, leading to increased potential for error.

This trend results in the cost of the isolators being added on to the cost of the cleanroom — the overall result is thus often a significant increase in both installation and running costs over conventional cleanroom operations.

A further apparent disadvantage of isolators has been the time and effort required for validation, especially where gassing systems are involved. It is unclear why isolators should be more difficult to validate than any other piece of pharmaceutical equipment. The technology is now quite mature and it should be possible to write and execute perfectly practical validation protocols. The issue of the validation of gassing processes is further discussed in Section 4, p66.

1.4 Ergonomics

While compliance may not be directly linked to ergonomic design, the potential for error will be greater if the operators are not comfortable or if they feel that the work is unsafe in some way. The overall layout of an isolator is likely to be governed by the process to be carried out within it, together with the equipment required for the process.

It may be useful to initially mock up the process on the open bench. If the work can be carried out and all of the equipment reached from a seated position, then the optimum solution will be handling by gloves. If, on the other hand, the work has to be carried out from a standing position, with greater reach and heavier lifting capability, then half-suits will probably be the better option. As a rough guide, it is possible to lift up to 5 kg over a radius of 500 mm in gloves while 15 kg can be lifted over a radius of 1,000 mm in a half-suit.

Some processes, such as freeze-dryers will demand direct interface with equipment, and here a full-size mock-up is almost essential. This will allow the designer to check that the operator can, for instance, open the dryer door fully, and then reach all the shelves inside for loading and unloading. It may be useful to make a video of the mock-up in operation for future reference.

As a general rule, operators should be able to reach all parts of the isolator not only for operation, but for cleaning and maintenance purposes. Shoulder rings for gauntlets or sleeves should be centred about 1150 mm from the floor and half-suit mounting plinths should be angled for entry, with the front edge at standard bench height, 900 mm from the floor.

2 OUTLINE OF THE TECHNOLOGY

2.1 Structures — Flexible and Rigid

Various materials have been used to construct isolators but now they broadly fall into two categories — 'flexible film' in which the main structure is clear PVC plastic film, or 'stainless steel', in which the main structure is sheet stainless steel with windows in clear rigid material. The very first pharmaceutical isolators, pioneered by the French company La Calhène in the 1980s, were based on enclosures developed to house research animals. These were formed from a complete 'bubble' of clear PVC plastic film, fabricated by radio frequency (RF) welding. This bubble was placed on a rigid work surface and constrained by a support frame of metal tubes.

This design was much improved in the late 1980s, notably by Cambridge Isolation Technology Ltd, with the application of a rigid stainless steel base tray. The clear PVC 'canopy' was then sealed around the rim of this tray while a more sophisticated ventilation system was fitted below the tray. Improved instrumentation and alarms also became the norm at this time.

Current flexible film isolators provide an excellent form of enclosure for processes such as sterility testing. The cost of these isolators is relatively low, the quality of the environment very high, and the clear all-round view is appreciated

The author at work in a flexible film half-suit isolator

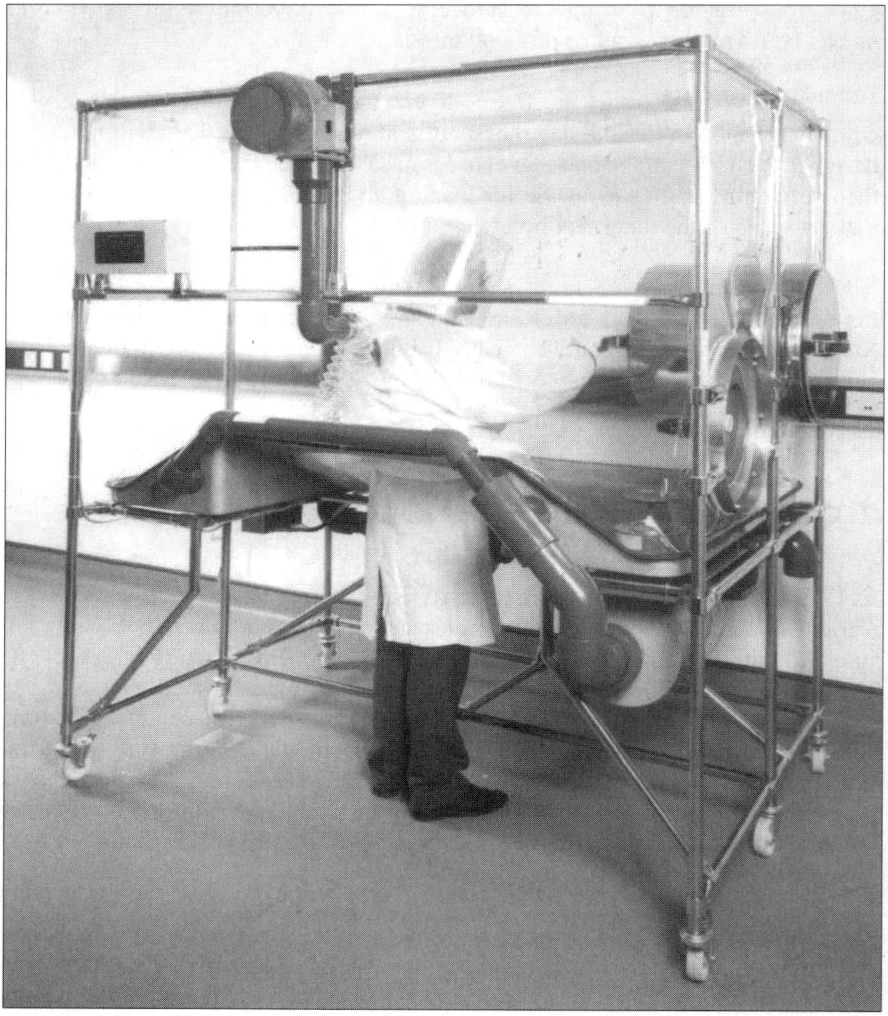

by operators and supervisors, especially where half-suit isolators are involved. These isolators can also be installed more easily than rigid structures, since they can be dismantled to pass through standard doorways. Flexible film tends to be perceived as a flimsy form of construction, but it is in fact remarkably durable. It will withstand over- and under-pressure far better than rigid structures, although it is fairly susceptible to penetration by sharp objects.

Flexible film canopies should be made of good quality clear PVC film from 0.50 mm to 1.00 mm in thickness. The RF welds should be 'blocked down', a secondary process which flattens out the weld and allows the weld line to seal where it passes over rigid parts of the structure, such as the base tray or shoulder rings. Base trays will normally be fabricated from stainless steel in much the same way as complete rigid isolators (described below).

Wherever possible the completed isolator structure should be mounted on castors so that it can be moved to allow for cleaning and maintenance.

More 'industrial' isolators, such as those for major filling lines, will be built from stainless steel and fitted with glass, acrylic or polycarbonate windows. Small isolators may use stainless steel 2–3 mm thick, while larger isolators may use 3–5 mm thick stainless steel. All edges should have a generous radius, not less than 20 mm, for cleaning, and corners should be ball formed at the same radius. The fabrication of the stainless steel must be crevice-free and fully dressed before finishing. The final finish should be around 0.60 micron roughness average (Ra) for internal surfaces and 1.2 micron for external surfaces. Ideally, the supplier should produce test data to show that the required Ra has been achieved at a number of test sites.

Isolator suppliers make much of the fact that they use stainless steel of 316L grade. This material should certainly be used for product contact equipment such as vessels but it is not necessary for isolator structures. Stainless steel 304 is perfectly adequate — it is easier to fabricate neatly, available in larger sheets which cuts down on welding and is less expensive than 316. Where extreme corrosion resistance is needed, for example where strong acids are present, one of the range of 'Hastelloy' alloys may be used, though this is a rare requirement.

2.2 Air Flow, Pressure and Filtration

2.2.1 Air Flow Regimes

It is the flow of filtered air through an isolator which purges the microbial and particle burden of the isolator interior and thus reduces both the potential for product or process contamination and the risk of non-compliance. The air flow is arguably a more critical parameter than the air pressure which so exercises many isolator pundits.

A stainless steel isolator used for cytotoxic compounding (Dabur)

There are three possible air flow regimes in isolators. Turbulent flow is the simplest version in which filtered air generally enters at one single point and leaves at another single point. The total air change (TAC) rate may vary from as little as 10 per hour up to 100 per hour, though the intention should be to maintain a turbulent flow pattern and thus hold particles in suspension, preventing them

Figure 2.1 Turbulent Flow Regime

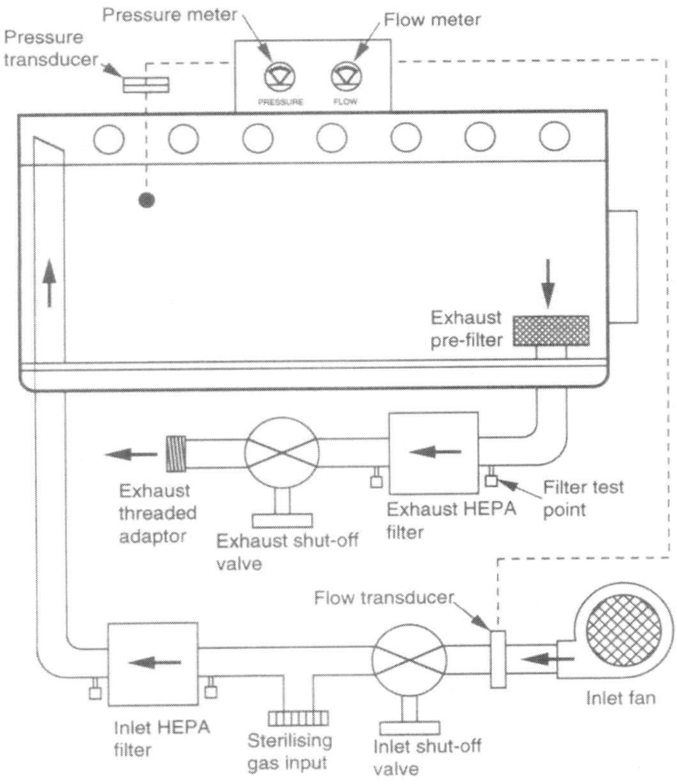

from settling out within the isolator. The system should be designed to purge all parts of the isolator volume, with no 'dead spots' or re-circulation zones, perhaps using an inlet air distribution device. Such air flow is suitable for operations where particle generation is likely to be low, as in sterility-testing, especially where enclosed methods such as 'Steritest' are used.

A major advantage of turbulent flow is low cost, because the filter arrangements are simple and the fan power required is small. The disadvantage is the relatively long residence time of any adventitious contamination within the

Figure 2.2 Unidirectional Down Flow Regime

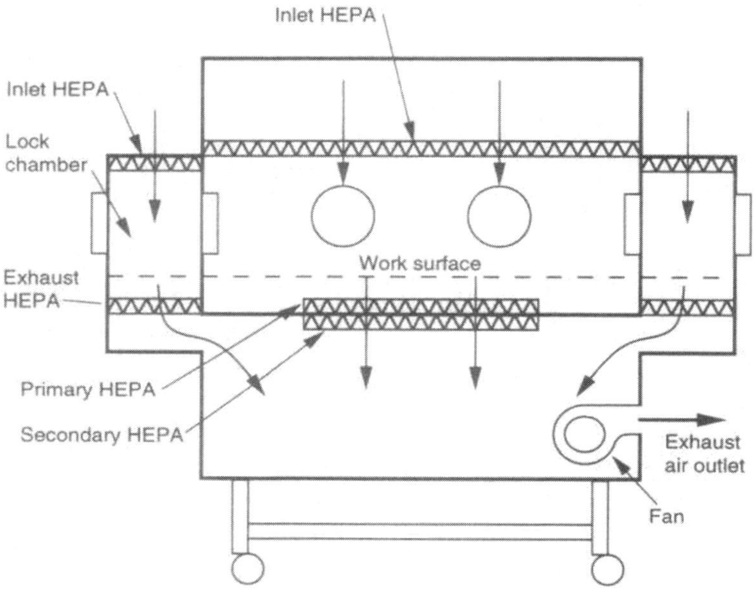

isolators. This ranges from a theoretical six minutes where the isolator has an airflow rate of 10 TAC per hour, down to 36 seconds for 100 TAC per hour. Turbulent flow is quite acceptable for processes such as sterility testing, but is less likely to provide compliant conditions where, for instance, a parenteral product is open to the isolator atmosphere. However, hospital pharmacy isolators with turbulent flow have proved perfectly workable, even when operated at negative pressure for cytotoxic compounding.

The unidirectional down flow regime (formerly referred to as laminar flow) produces a very short residence time for particles, as little as 0.40 seconds for an isolator chamber 1 metre high. The standard target for down-flow velocity will be 0.36 m/s to 0.54 m/s (EC GMP guidance value), although some users report very good results with much lower velocities, as low as 0.05 m/s. Low down flow velocity may be useful where precise weighing operations are involved.

Figure 2.3 Mixed Flow Regime

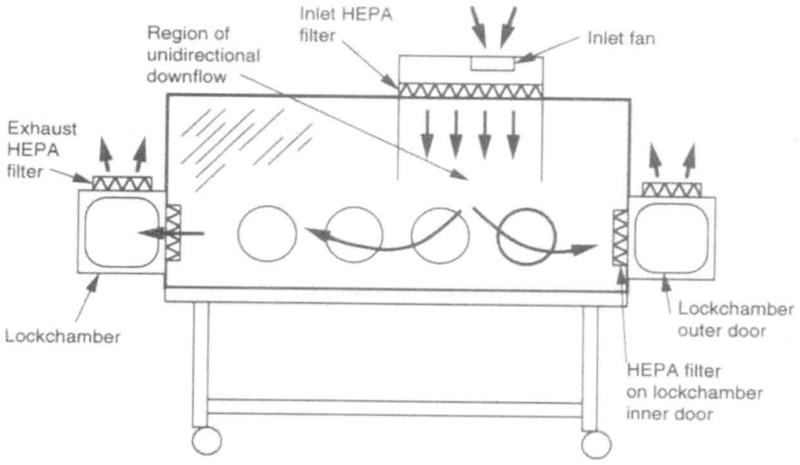

Isolators with unidirectional air flow regimes are significantly more expensive, both to build and maintain, since large areas of panel HEPA filters must be fitted, together with the associated plenum chambers above and large fans to drive them. Arrangements must be made to collect the exhaust air with minimum disruption to the flow. Perforated floors are not practical and so exhaust grills are normally fitted along the front and back edges of the base tray, leading to an exhaust plenum chamber below. Alternatively, double layer windows have been used to provide a return air duct in unidirectional isolators. These more complex ventilation systems require quite extensive engineering, demand more power, more instrumentation and can produce more noise.

Isolators are, almost by definition, quite limited in size, and contain sleeves, half-suits and various pieces of process equipment. Thus unidirectional down flow is actually broken up into turbulent flow quite quickly after leaving the filter face. Their virtue may lie less in the laminarity of the air flow than in the very high air change rates provided.

Mixed flow regimes provide something of a compromise. Here a region of local unidirectional down flow is provided within a larger volume of turbulent flow. Critical operations, perhaps where vials of a parenteral product are open, can be carried out in the unidirectional flow zone while stock holding and handling are carried out in the turbulent zone. This system has been used successfully in a series of hospital pharmacy isolators.

As a final consideration, whatever air flow regime is chosen, its impact on the heating, ventilation and air conditioning (HVAC) system of the background room has to be taken into account. In small turbulent flow isolators, the impact may be very limited. In large unidirectional flow isolators the requirement for air flow may be so great as to require its own dedicated AHU (air handling unit).

2.2.2 Pressure regimes

Isolators can be run at a slight positive pressure with respect to atmospheric pressure, or at a slight negative pressure. Where the primary function of the isolator is to maintain the purity of the product, as in parenteral preparations, then the isolator will certainly be run at positive pressure. Thus any net leakage will be out of the isolator. Where the primary concern is operator safety, perhaps in handling cytotoxic materials, the isolator will be run at negative pressure and any leakage will be inwards, away from the operator.

The choice of isolator pressure regime becomes the subject of debate where both the product and the operators must be protected — the prime example of this dilemma is in the handling of the modern range of cytotoxic (anti-cancer) drugs. These products are often administered as intravenous infusions and must therefore be sterile. They are however toxic, even at quite low levels, if the exposure is continuous, as could be the case for professional operators.

This problem was the subject of a study by the Medicines and Healthcare Products Regulatory Agency (MHRA) and the Health and Safety Executive (HSE) in the UK (HSE/MHRA, 2003). The resulting paper revealed that the operating pressure of an isolator is not a major factor in defining either the product quality or the operator safety. Other issues such as sanitisation, transfer and general handling procedures have a far greater influence. The conclusion is that cytotoxic isolators may be operated at either positive or negative pressure, but whichever regime is chosen, it must be backed up by a clear rationale. In practice, most industry users run their cytotoxic isolators at positive pressure unless dry powders are to be handled. By contrast, hospital pharmacies tend to run at negative pressure for cytotoxic compounding and dispensing. Such decisions may be based on politics rather than science.

2.2.3 Air filtration

Air filtration is a primary function of the isolator. Where aseptic work is involved, a biological filter is needed on the inlet air. Where toxic work is involved, a fine filter is needed on the exhaust air. In practice, filters are fitted to both inlet and exhaust air, on both aseptic and toxic isolators. This then provides a definite, physical barrier for particles through inlet or exhaust and in many cases also helps to maintain the pressure regime.

Isolator filters are almost invariably HEPA or ultra-high efficiency particulate air (ULPA) filters. These are built up from very fine-grained glass-fibre paper, pleated to maximise the surface area. They may be in the form of panels or in some cases, canisters. HEPA filters are not simply sieves but operate with a series of physical processes outside the scope of this chapter (Thomas, 1994). The efficiency of the HEPA filter varies with particle size and it should be noted that most have a minimum efficiency at around 0.30 micron particle size. This is known as the most penetrating particle size (MPPS) of the filter and is a factor in filter testing. Perhaps curiously, both above and also *below* this size, the efficiency of the filter rises.

The performance of HEPA and ULPA filters is gauged by the efficiency of filtration at a given size (or sizes) of particles, commonly at 0.30 micron, the MPPS. HEPA filters are usually defined as having an efficiency of 99.997% at 0.30 micron, ie 0.003% penetration. ULPA filters are usually defined as having 99.9999% efficiency, ie 0.0001% penetration. HEPA filters are normally quite sufficient for pharmaceutical processes and only the semi-conductor industry resorts to ULPA filters.

Small turbulent flow isolators often use cartridge HEPA filters, mounted in plastic housings. These are around 250 mm in diameter and 250 mm long, and are mounted directly in the ductwork. The filter element may be changeable or the whole unit may be disposable.

Larger turbulent flow isolators and all unidirectional downflow isolators will have panel HEPA filters, mounted to form part or virtually all of the ceiling of the working area. Some care needs to be taken in the design of such filter mountings. First, the filters should be fitted 'as the duct' and not 'in the duct', to minimise the potential for contamination through leakage of the seals (see Figure 2.4)

Second, the clamping arrangements must produce a good seal and may allow up to 50% compression of the filter seal, but no more. The compression force needs to be even all the way around the seal surface, with no distortion of the filter frame. In unidirectional down flow isolators, consideration could be given to gel

Double exhaust filters fitted to a flexible film isolator. The primary exhaust filter is safe-change by removal into the body of the isolator

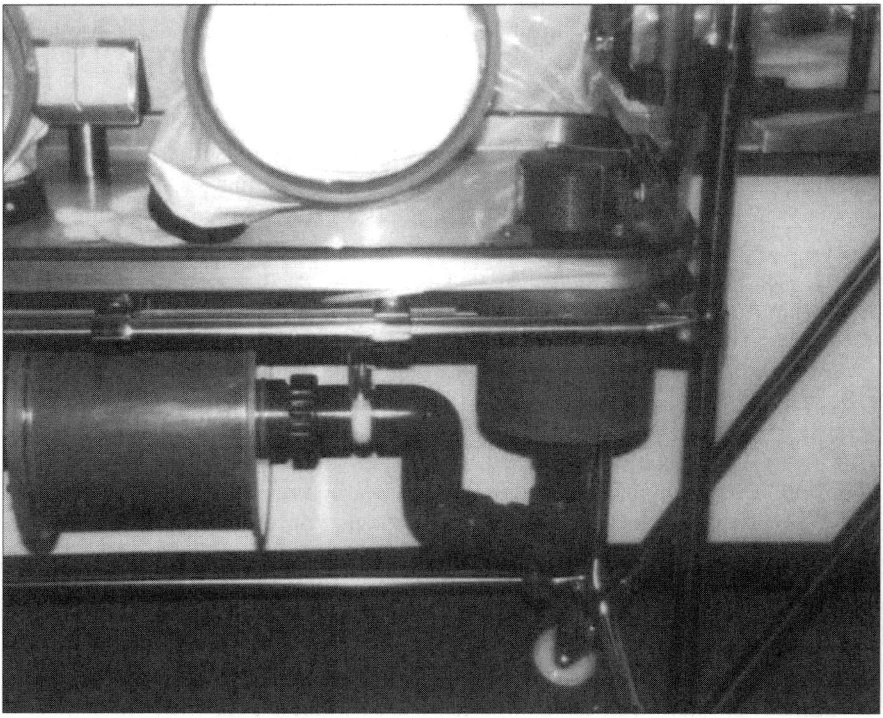

seals, which effectively eliminate leakage. With large panel filters, the plenum chamber above should be deep enough to ensure that the same pressure is applied across the entire filter face.

Third, the designer has to think about the commissioning and routine testing of all the filters, supply and exhaust, whether single or double. Each filter requires three test points.

- A point to introduce the test aerosol, usually oil smoke. This has to be far enough upstream of the filter face to allow full mixing of the smoke in the air flow.

Figure 2.4 HEPA filter mounting

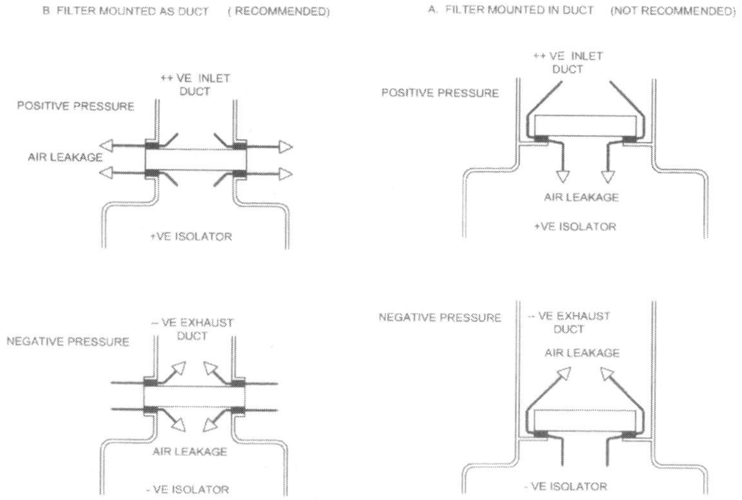

- A point to sample the concentration of the test smoke immediately upstream of the filter face. With large panel filters, some form of grid sampling device may be required to produce a representative sample.

- A point to sample to air flow downstream from the filter face. With unidirectional down flow isolators, this means scanning across the face of the inlet filter. For other filters this may mean a volumetric sample, well downstream of the filter face, again to allow full mixing of the air flow.

A single inlet and exhaust filter may be fitted to the isolator but in the UK the MHRA prefers double inlet HEPA filters, for added security in aseptic applications. For toxic operations, double exhaust filters may be specified with the primary filter being changed into the isolator (safe change), while the exhaust ductwork is still protected by the secondary filter.

As a last consideration, the filter test points should be easily accessible and the filters themselves should be changeable without major rebuilding works. The HEPA filters are critical for isolator compliance so time and money spent on good engineering of the mounting arrangements and test points is well invested.

2.3 Handling Methods

Excluding full air suits and robots at this stage, there are basically two methods of handling inside isolators — gauntlets (or glove-sleeve assemblies) and half-suits. The choice depends on the nature of the work to be carried out, as mentioned in Section 1, p42.

Single-piece gauntlets made from neoprene or Hypalon® are generally favoured in toxic applications, firmly secured to a shoulder ring sealed to the isolator front window. Shoulder rings may be circular or oval, to allow greater access without obscuring the view through the adjacent window. Gauntlets, especially in Hypalon®, are quite expensive and are therefore changed relatively infrequently. This is less than ideal for compliance, since the gauntlets are perhaps the weakest link in the barrier and place the main source of contamination (the operator) close to the process. The 'concertina' type gauntlets favoured by the nuclear industry are definitely less compliant than ordinary versions as the folds are difficult to clean and sanitise.

Sleeves with separate cuffs to carry gloves are common in aseptic applications. These should be comfortable on the operator side and smooth on the isolator side, thus double-layer sleeves may be preferred. An added benefit here is that if either layer becomes punctured, air is admitted between the layers and the failure becomes evident. Sleeves will normally have O-rings welded into each end to seal them down onto their respective shoulder and cuff rings. Users should examine this weld carefully to ensure the enclosure of the O-ring is not wrinkled and that the transverse weld line is fully blocked down (flattened out) so that the ends of the sleeve seal properly.

There are various types of cuff assembly, some of which are usefully designed to allow glove change without break of containment. The gloves are not only the weakest link in most isolator systems, but also place one of the most contaminated items (the operator) close to the work. Thus the choice of gloves presents a dilemma between the dexterity and 'feel' for the operator, and the strength or robustness of the glove. Ideally, the glove material should be as thick as practical for the work in hand. The gloves must fit the operator well and ambidextrous gloves should be avoided if possible. The material should resist the products, solvents and cleaning materials present in the isolator. Special consideration may be given to handling products such as cytotoxics which are known to diffuse through glove materials (Thomas and Fenton-May, 1987). In practice however, frequent glove change overcomes diffusion problems. Double-gloving is often advised, especially in cytotoxic work, which aids compliance in both safety and GMP.

Half-suits are a very useful device where heavier and larger objects are to be manipulated in the isolator, or where greater reach is needed. The suit covers the body from the waist upwards and is mounted in a 'pulpit' or plinth, generally on the edge of the isolator. This plinth is angled upwards at the back to allow easy entry for the operator and is oval in shape to give more reach to the sides. The industry-standard half-suit oval is 800 mm from left to right and 500 mm from front to back. The suit is fitted with an air supply, commonly fed to the neck region and the cuffs, to give a pleasant air flow over the face, arms and body. Suits should be as light and comfortable as possible for extended operation, and users may want to try out several different makes before choosing the best for their application.

Users will also want to give very careful consideration to the manner in which the suit is supported when unoccupied. This support should extend the suit fully for gassing (the so-called 'don't shoot' position), but allow simple and easy entry and exit. Complex suspension systems with elastic cords and catches, or loose hook systems, must be avoided. Pull-out or fold-down steps may be fitted under half-suits to help shorter operators. Half-suits are usually made out of the same sort of materials as sleeves and are surprisingly comfortable and practical when designed and fitted correctly.

2.4 Control, Instrumentation, Alarms and Data Recording

Control and instrumentation are important aspects of compliance for isolators. Most modern isolators have active electronic control of the canopy pressure and the air flow rate (or down flow velocity). These parameters are adjusted by control of the inlet fan and in many cases by the exhaust fan also. Thus the operator may set a flow rate which is then maintained by the exhaust fan. He may also set a canopy pressure which is then maintained by the inlet fan. Digital display of both these parameters is usually offered, together with alarm systems. Air flow transducers come in a variety of forms but the orifice plate may give the most accurate results, despite the increased flow resistance which it creates.

A good isolator will have user-set alarm limits on both high and low excursion for both air flow and canopy pressure. The alarms will be visual (eg a flashing beacon), audible (eg a sounder) and remote (eg connection to the Building Management System (BMS) or Facility Management System (FMS)). Ideally, any alarm should clearly identify the nature of the problem, eg 'low canopy pressure'. Alarms must be latched — this means that once they are triggered, they can only be re-set by active operator intervention. Without this feature, the isolator may go outside limits for a period, perhaps overnight, and then return itself to normal, with no indication that it has run in a non-compliant condition.

A typical half-suit in use. Note the holes in the inner layer which deliver air to the operator

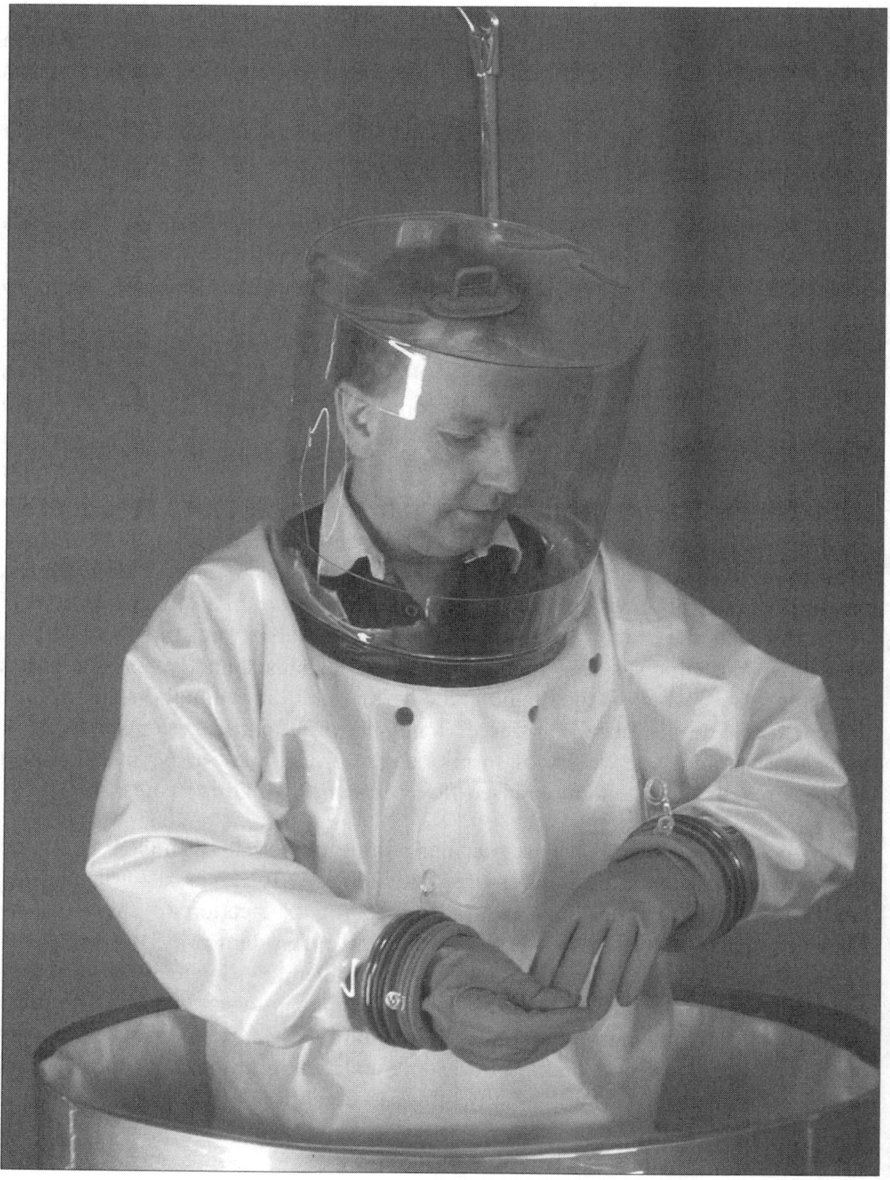

Isolators are frequently fitted with magnehelic gauges which read the differential pressure (DP) across each HEPA filter. The logic here is that operators can quickly see if the filter is blinding and needs changing. However, isolators generally operate in relatively clean conditions and the HEPAs are likely to run for a number of years before they show any appreciable increase in DP. Beyond this, provided that the canopy pressure and air flow rate are within limits, the HEPA filters are operating satisfactorily. Thus such gauges are likely to be more confusing than useful to operators.

Another compliance aspect of control and alarm systems has in the past been the issue of poor labelling. All isolator controls, warning lights and displays must be clearly and unambiguously labelled. Furthermore, this labelling must match precisely with the wording of the Operation and Maintenance (O&M) Manual. Labelling should extend also to the component parts of the ventilation system, especially any sanitising gas connection points, stop-valves and filter test ports.

Data recording is useful in assisting with compliance. Records of the isolator performance, usually in terms of flow and pressure, can be of great benefit if any query arises with product quality or process performance. Modern digital paperless chart recorders make this relatively simple and cheap.

If the isolator provides some specific condition such as low relative humidity (RH) or nitrogen atmosphere, then instruments should be provided to monitor this condition, probably linked through to the alarm system. All instruments must be regularly calibrated back to appropriate standards.

3 TRANSFER METHODS

3.1 Doors, Mouseholes Lockchambers, Rapid Gassing Ports and Product Passout Ports

Integral to isolator operation are the transfer of materials into an isolator and the removal of finished products and waste. Provided HEPA filters are operating correctly, the gloves are sound and the sanitisation process has worked as intended, the only real risk of contamination in an aseptic isolator lies with the transfer processes. The same is effectively true for containment isolators; it is transfer that is most likely to lead to escape of material. Thus if compliance is your target, then time spent specifying and validating the various transfer processes is time well spent. Transfer devices are well-described in the 'Yellow Guide' (Midcalf et al, 2004), which denotes them from 'A1' through to 'F' in a series which roughly increases in security of transfer thus:

- A1 simple door
- A2 mousehole
- B1 simple lockchamber
- B2 gassable lockchamber
- C1 lockchamber with filtered air exhaust
- C2 lockchamber with filtered air inlet
- D lockchamber with filtered air inlet and exhaust
- E gassable lockchamber with filtered air inlet and exhaust
- F double-door transfer ports (see section 3.2, p60).

The most basic transfer system is the simple door (A1). This is opened, materials are placed in the isolator, the door is closed and the process carried out, including sporicidal gassing as appropriate. At the end of the process, the door is opened and the finished products and waste are removed, ready for another cycle. This scheme is ideal for tasks such as small-scale sterility testing, since both the equipment and the operation are simple and the risk of contamination is minimal. There are no major compliance issues with simple doors but none the less careful design is required to produce a door that is both easy to use and seals reliably on closure.

In many cases, however, the main isolator has to continue in operation while materials are transferred in and products are transferred out. This then means that suitable transfer devices must be fitted, and a range of devices are on offer to the user. The first of these is the so-called 'mousehole' (A2), which takes the form of a minimum-sized aperture in the wall of the isolator through which items such as filled vials can pass continuously. To reduce the potential for back-flow of air (in positive pressure isolators), the mousehole is often covered by an area of unidirectional downflow filtered air, producing the 'dynamic mousehole'. Of course, continuous conveyor belts cannot pass through mouseholes and so the threshold of the mousehole is always a dead-plate. Consideration also has to be given to sealing the mousehole appropriately during sporicidal gassing. The validation of mouseholes presents some interesting problems, with smoke pattern studies providing an initial check, perhaps followed by environmental monitoring, rather than applying a direct challenge.

Type D lockchamber fitted to a flexible film isolator

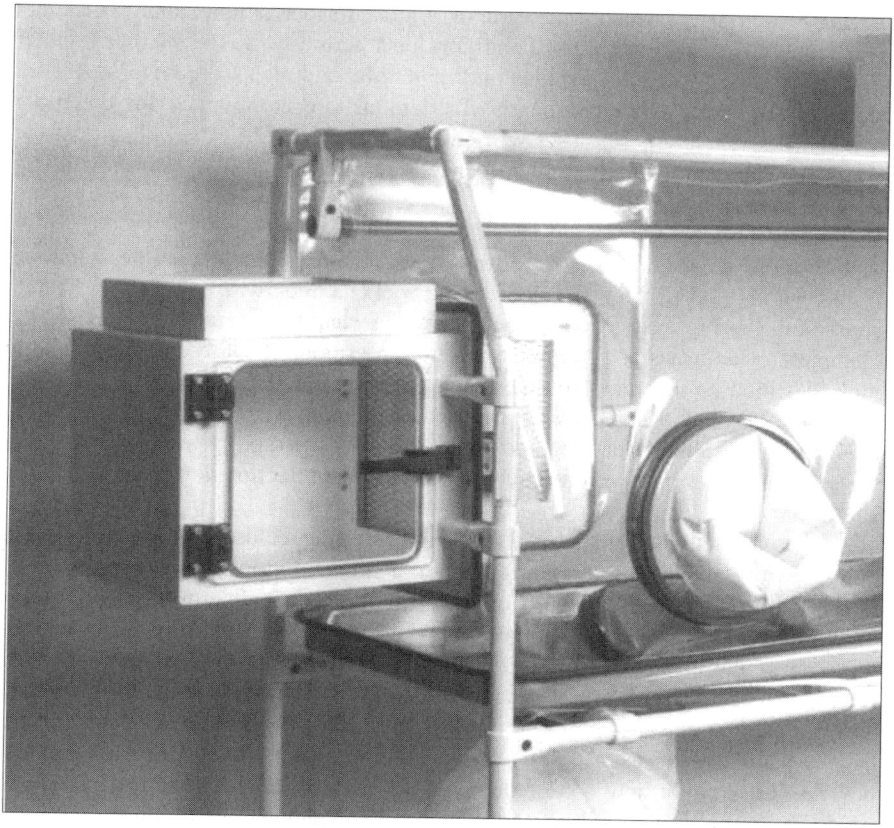

The next series of transfer devices are lockchambers, also (though less accurately) termed pass-through hatches or passboxes, which come in a variety of forms from B1 to E as listed above. While all of these have been used at some time, in practice the type D is most commonly used, particularly on hospital pharmacy isolators. These lockchambers frequently have timed interlocks so that the volume of contaminated air introduced when the outer door is open is purged by incoming filtered air, before the inner door is allowed to open. Combined with the use of sporicidal, or at least bactericidal, sprays these lockchambers give good results in hospital pharmacy applications.

The Type E lockchamber may be a small structure — perhaps 300mm in each dimension, but can equally well be another complete isolator. Thus a typical sterility testing suite consists of a two-glove isolator loaded via a simple door and linked to a four-glove working isolator with a further simple door. The main isolator is maintained in aseptic condition for a period while the smaller transfer isolator is used to deliver samples and materials, and also to remove processed samples and waste more frequently. To demonstrate compliance for the more complex lockchambers, the user will need to test the HEPA filters, leak test the doors and the complete structure and check the interlock system, along with any instrumentation and alarms.

A more recent device covered by the Type E designation, is the so-called 'rapid gassing port'. This is effectively a lockchamber with its own dedicated sporicidal gassing system designed to gas the chamber as rapidly as possible. Complete cycle times as brief as 15 minutes are quoted for these devices, which look destined to become commonplace aseptic transfer ports in the future. However, they are less likely to be used in applications such as the hospital pharmacy due to their very high cost. Validation of the rapid gassing port will be the same as for any gassed enclosure, as discussed in Section 4, p66.

The product passout port is a further transfer device used for sealed product delivery from an isolator, though it does not find an obvious place in the transfer device lettering system. This port consists essentially of a length of polyethylene continuous plastic bag or 'layflat' tubing which is 'reefed' onto a suitable ring on the wall of the isolator. Individual items are passed out of the isolator into the continuous bag and then a heat bar is used to seal across the liner behind the item. The bag is then cut to release the item which is thus held in a sealed pouch and can be removed for use elsewhere. The port may include a door and various devices for fitting a new continuous bag when all of the previous bag has been used. The layflat tubing will normally be gamma irradiated prior to use.

3.2 Rapid Transfer Ports

Rapid transfer ports were first developed for the nuclear industry by the French company La Calhène SA, with application in the pharmaceutical industry following on in the early 1980s. They use the 'double-door' principle to make an immediate contained transfer from a container to an isolator. The port consists of six basic elements.

- container flange
- container lid

- container lid seal
- port flange
- port door
- port door seal.

In operation, the container with its sealed lid is 'docked' onto the port with its sealed door. The docking process simultaneously:

- locks the container flange to the port flange
- releases the lid from the container
- locks the container lid onto the port door.

The combined container lid and port door can then be opened as a unit into the isolator, and the transfer can be made. The container lid seal and the port door seal are 'arrowhead' shaped, with the points of the two arrows coming together to minimise the potentially contaminated common contact area (see Figures 2.5–2.7).

In general the port assembly is fitted onto a fixed isolator — also termed the 'female' or 'alpha' part of the system. Likewise the container assembly is normally fitted to a mobile structure and is termed the 'male' or 'beta' part.

The RTP can be used in a variety of ways. A good example of usage is the delivery of rubber stoppers to a filling line in an aseptic isolator. The stoppers are loaded into either a stainless steel or a 'tyvec' RTP container and autoclaved. The container is then docked onto the filling line and the sterile stoppers are fed to the hopper inside the isolator. A more recent version of this has a special RTP fitted directly onto the door of the autoclave and container bags are filled directly from the autoclave, with vial or syringe stoppers.

In another example, the container flange and lid assembly are fitted onto the wall of a complete mobile isolator, so that this transfer isolator can be docked onto another fixed isolator. Thus trays of empty vials may be off-loaded from a depyrogenation tunnel into the transfer isolator, which is then moved over to dock with a filling line isolator to deliver the vials.

Most RTPs are opened manually from inside the isolator but new developments allow for outside opening. This then means that the port can be

Figure 2.5 A Cross-Sectional Drawing of a Rapid Transfer Port and Container

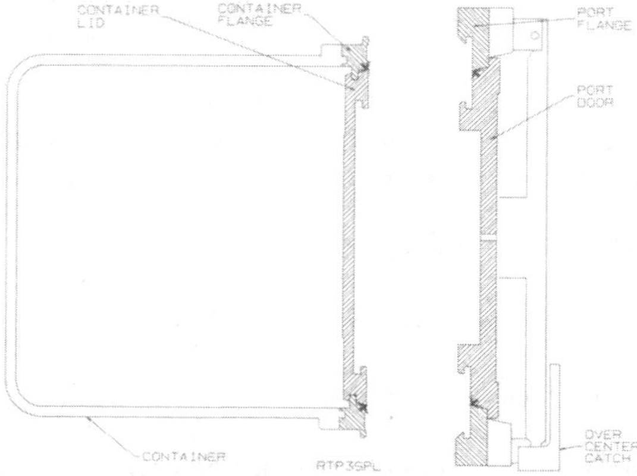

This figure shows the container and the port separate, prior to docking the two together

Figure 2.6 A Rapid Transfer Container in Place after docking with the Port

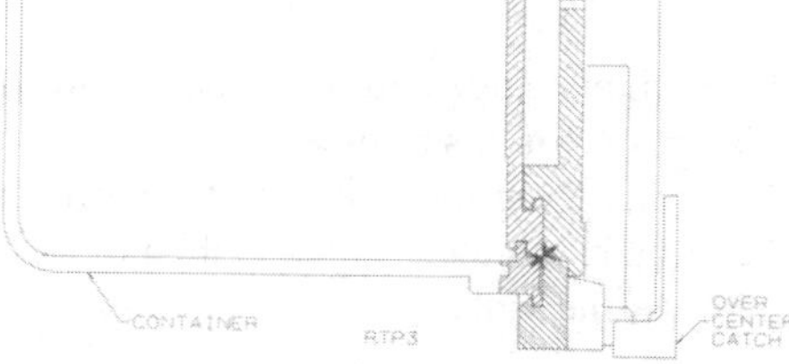

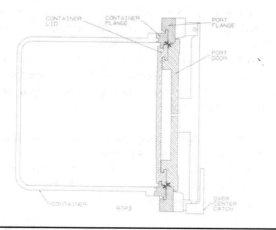

Note how the arrowhead seals on the container and the port meet point to point

Figure 2.7 An Open Port Door

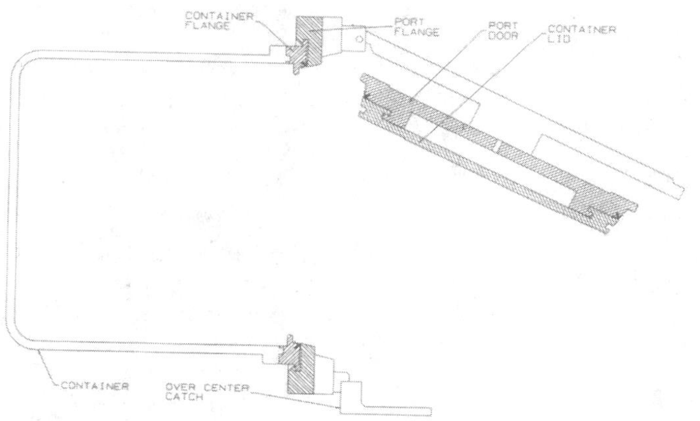

In this drawing, the port door has been opened, bringing with it the container lid, their contaminated surfaces being sealed gas-tight. Transfer can now take place without further sterilisation or decontamination

A Transfer Isolator Fitted with a Container Flange Assembly

An IDC reactor charging port. In this graphic, the port is fitted with a pressure-lid ready for CIP/SIP operation

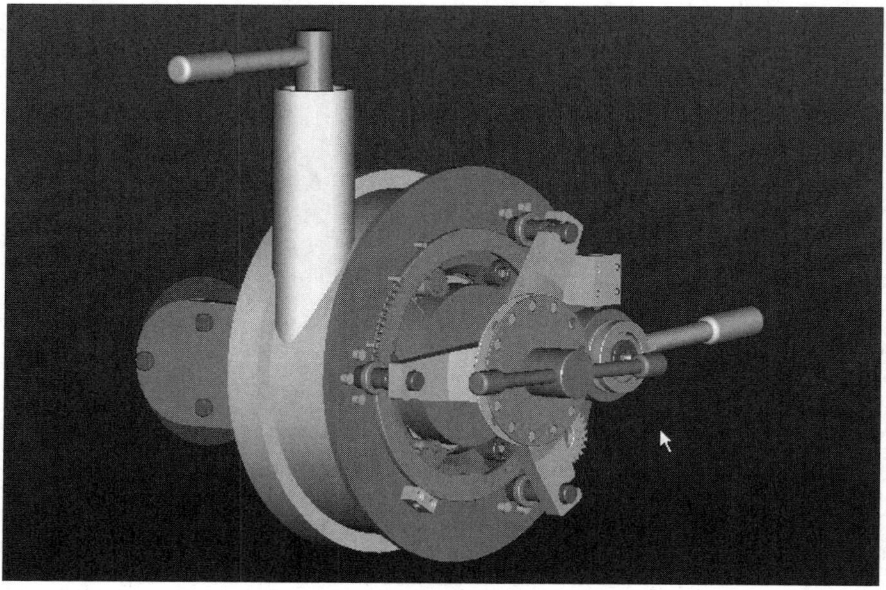

fitted onto vessels such as reactors and thus used for their direct aseptic or toxic charging. Such ports are also designed to withstand the pressure of clean-in-place/sterilise-in-place (CIP/SIP) cycles. Powder transfer tests on these ports have demonstrated containment levels into single figures of nano-grams.

Concern has been expressed over the so-called 'ring of confidence' presented by all RTPs. This is the area at which the two seals meet point-to-point and where cross-contamination could occur, even in the best engineered of ports. Users may feel that newly-installed RTPs should therefore be individually validated by some form of direct challenge. In practice however, these ports have been well tried and tested over the years and, if operated and maintained correctly, perform as if each transfer were truly aseptic.

Naturally sensible precautions should be taken to minimise the challenge presented to the port. For example, aseptic ports should be operated in a cleanroom of at least ISO Class 8 and preferably Class 7, while the face of the RTC and port may be wiped with a sporicidal agent before each docking action takes place. Taken overall however, the RTP system works well as an isolator transfer devices. A version of the port is produced by CRL in the USA, under licence from La Calhène, which has an electrical heating element designed to sterilise the ring during the docking process. Those RTPs fitted with mechanical interlock systems are preferred, so that the port door cannot be opened if no container is in place, and the container cannot be removed if the port door is still open.

Final note — the correct pronunciation of 'La Calhène' in both French and English is la-kal-enn (The 'h' is not pronounced and the first 'e' carries a grave accent not an acute accent)

3.3 Direct Interface and Utilities

Isolators are often required to interface with process equipment such as:

- vial, ampoule and syringe filling machines
- depyrogenation tunnels
- autoclaves
- freeze driers
- E-beam chambers
- reactors
- blenders
- filter driers.

The detailed engineering of the interface can present some serious challenges and careful design is required. If the process equipment and the isolator are provided by different manufacturers, as is often the case, then close liaison will be needed to produce a good working interface.

The doors of autoclaves and freeze driers present a problem in aseptic applications. If these machines are equipped with sliding doors, consideration has to be given to the means of effectively gas sanitising the volume which houses the open door, and then validating this process. This issue is simplified if the door can be made to open out into the main volume of the isolator.

The union of an isolator with a filling machine can also be problematic. The question of mechanical fit has been overcome in the past when the filling machine maker has supplied a dummy top plate to the isolator manufacturer. If the isolator is then built to fit this dummy plate, there is good assurance that it will fit onto the final filling machine. Leak testing of these combined units can lead to disagreement, with each supplier blaming the other if the complete assembly does not meet the URS-stated maximum leak rate. It may help if each unit is leak tested and made compliant separately, after which the completed assembly may be tested with some degree of confidence.

Various utilities may be needed for the process inside the isolator, in particular mains electricity supply. Cleanroom power sockets are frequently fitted inside isolators but even these generally contain voids and crevices that are hard to clean, have poor sanitising gas penetration and may lead to leakage. It can be simpler and more effective to lead power cables through compression glands in the wall of the isolator, though it should be noted that multi-strand cables can, themselves, act as a significant leak path.

Care must be exercised where compressed air, gas or vacuum are fitted to an isolator. If either of these has a major failure, massive over- or under-pressure of the isolator will occur with serious consequences. If either compressed air or vacuum leak by a small amount while the isolator is not running and is sealed, again over-pressure or under-pressure may result, significantly damaging the isolator structure.

4 SPORICIDAL GASSING

4.1 Gas Generators — Introduction

By their nature, unlike cleanrooms, isolators bring their walls, handling means and transfer devices relatively close to the process. It therefore makes sense, in aseptic applications, to look for means to 'sterilise' the internal surfaces of the isolator and thus further reduce the potential for contamination. The fact that isolators are small sealed enclosures suggests that some form of sterilisation by gassing would be appropriate and convenient. Thus various types of gas generator have been developed to treat the inside of isolators.

Two points should be noted at this stage:

- Gassing is only the final 'polishing' stage of the complete bio-decontamination process. The process starts with gross cleaning (eg sweeping up the broken glass and mopping up spills), followed by detailed cleaning (eg systematic wipe-down with a liquid sporicidal agent) and finally finishes with gassing.

- The gassing process cannot be termed 'sterilisation' since only heat (wet or dry) and gamma-irradiation produce a truly sterile condition. Gassing is a process that can reduce the viable micro-flora very greatly, but does not guarantee to inactivate every viable microorganism. For this reason, the process is generally referred to as 'sporicidal gassing'. It does, however, approach total inactivation when correctly carried out. Note also the use of the term 'sporicidal' to embrace the kill of not only the vegetative forms of the micro-flora but also their more resistant spores.

4.2 Gas Generators — Internal Generation, Open Loop and Closed Loop

The very simplest form of gas generator is the formaldehyde 'kettle' which boils off an appropriate volume of formalin solution inside the isolator. The vapour is allowed to dwell for a time and is then blown away to atmosphere. The process is still used in Class 3 biological safety cabinets but the toxicity of the material, its relative inefficiency and uncontrolled nature make it virtually unused in other isolator applications.

The next form of gas generator is the open-loop type, which appeared around 25 years ago. This generated formaldehyde vapour originally but went on the use peracetic acid ($CH_3COO.OH$), hydrogen peroxide (H_2O_2) and mixtures of the two ('Citanox'). The open-loop generators blow the gas into the isolator, usually ahead of the inlet HEPA filters, through the volume of the isolator, through the exhaust HEPA filters and then out to atmosphere. These generators are simple and inexpensive but have the disadvantage of needing a duct to atmosphere. This can be difficult to engineer and causes environmental concerns. Thus the open-loop generators have fallen out of favour generally. Examples of open-loop gas generators include the La Calhène 'Sterivap' and the CIT (MDH — now BioQuell) 'Citomat'.

Current closed-loop gas generators produce vapour phase hydrogen peroxide (VPHP) by flash-evaporation of a 35% aqueous solution. The vapour is passed through the isolator and back to the generator where it is catalytically broken down to oxygen and water vapour. Thus the system is more-or-less totally closed so that there are no emissions, no requirement for a duct and no environmental concerns. Furthermore, the process is automatic and up to a point, self-monitoring, in terms of flow rates, temperatures and humidity.

The major advantage of using hydrogen peroxide as opposed to other gaseous or vapour phase agents (a vapour is the gaseous phase of a substance that is normally a liquid at room temperature) is that it breaks down to form oxygen and water vapour which are, of course, entirely innocuous. The breakdown can be accelerated by using a catalyser but will happen spontaneously over time. No residues are left by the process to affect either product or operator. A further advantage of hydrogen peroxide is that, while it is a powerful oxidising agent, it appears to cause little corrosion damage to the isolator and its equipment.

The VPHP generator may be a free-standing device connected by hoses to the isolator for the gassing cycle (eg Steris and BioQuell). Alternatively, the gassing system may be built into the isolator structure (eg SKAN). At the time of writing, the sporicidal gas generator market is largely dominated by the two manufacturers, Steris in the USA and more recently, BioQuell in the UK. Both work on a gassing cycle with the following stages.

1. De-humidification of the air in the isolator, down to 40% or less. This stage has been termed ‘pre-conditioning’ by Steris.

2. Injection of VPHP to raise the concentration in the isolator to around 2 mg/l (about 2,500 ppm by weight) — termed ‘conditioning’.

3. Maintenance of the VPHP concentration for a period of time. This has been termed both the ‘sterilisation’ and ‘gassing’ stages.

4. Reduction of the VPHP concentration to an acceptable level for isolator operation, usually less than 1 ppm — termed ‘aeration’.

The two use similar technology but with some differences. The Steris generator uses a chemical bed to dehumidify the air while the BioQuell generator uses refrigeration. The advantage of the latter is that it can operate continuously while the chemical drier needs periodic regeneration. The Steris generator continuously introduces VPHP into the circulating air and continuously breaks it down, thus injecting fresh vapour all the time. The BioQuell generator re-circulates a primary injection of VPHP into the system and then only ‘tops up’ with VPHP to maintain concentration. The advantage here is much lower use of hydrogen peroxide solution. Incidentally, vapour phase hydrogen peroxide is often referred to as ‘VHP’ however VHP® is the registered trade mark of the Steris Corporation.

In terms of compliance, the ideal control method for VPHP systems would use an instrument which directly measures the concentration of VPHP in the isolator and adjusts the injection rate accordingly, throughout the entire cycle. While there are instruments which can, in theory, measure the VPHP concentration directly, at the time of writing none of these are a practical proposition for cycle control. The current generators therefore rely on parametric control, monitoring and recording temperature and flow rates to hold the validated conditions of the cycle. Some operators do however, use independent VPHP instruments (eg the Dräger Polytron) to give an additional level of cycle record and have achieved useful results.

VPHP can be introduced into the isolator by various routes. Originally, the vapour was injected ahead of the inlet HEPA filter, passed through the isolator and removed downstream of the exhaust HEPA filter. This certainly treats the entire system including the filters and ductwork, but has the disadvantage that vapour is often lost to the inlet filter. This results in a significantly increased cycle time and so direct injection into the isolator volume has been used more recently. This leaves a query over the treatment of the HEPA filters and ductwork, however. it can be demonstrated that diffusion takes the vapour through the filters in sufficient concentration to give the required kill (see Section 4.4, p70).

In order to achieve good kill throughout the isolator, it has been shown that VPHP should be circulated quite energetically within the isolator volume. One or more axial fans can be mounted in the isolator or a delivery nozzle rotating in two planes (BioQuell patent) may be used.

4.3 Vapour Phase Hydrogen Peroxide — The Wet or Dry Issue

Some debate, acrimonious at times, has developed over whether the VPHP cycle is a wet or a dry process. Broadly-speaking, the two views are:

- the process is dry, takes place essentially in the vapour phase and ideally, no visible condensation should take place
- the process is wet, takes place in the liquid phase and visible condensation is acceptable, indeed even desirable.

The debate is interesting, has helped to gain insight into the VPHP process, but is essentially a commercial argument between two manufacturers, not a scientific one. It is simply a question of how the cycle is controlled, specifically the rate of VPHP injection. There are three possible scenarios:

- true vapour phase
- 'micro-condensation'
- true liquid phase ('macro-condensation').

It has been shown that the sporicidal kill process is very slow in the true vapour phase (Watling and Parks, 2004). However fast kill has been observed with no apparent, visible condensation taking place and it therefore appeared that the process was dry. Deeper investigation however, revealed that the sort of vapour phase concentration apparently produced could not physically exist. The injection rates used would deliver a concentration of over 2,000 ppm but calculation showed that only 200 ppm could be sustained in the vapour phase. What happened to the excess VPHP? The answer came in the form of a series of electron micrographs which showed the formation of small droplets of liquid on the surfaces of the isolator. These droplets measured a few microns in diameter and were thus not visible to the naked eye. Beyond this, the droplets appeared to form preferentially on the surface of microorganisms; furthermore, calculation showed that the concentration of hydrogen peroxide in the initially-forming droplets could be very high. This would account for the fast kill in what appeared to be dry conditions.

This discovery is important and gives vital insight into the conditions which must be set up and maintained within the isolator to produce a fast and effective sporicidal gassing cycle.

There is still some debate as to the usefulness of visible macro-condensation. This condition certainly produces a fast kill but may have some disadvantages. First, the accumulated liquid may take time to evaporate and perhaps penetrates more deeply into surfaces such as plastics, increasing the aeration time. Second, the liquid may cause more extensive corrosion damage to the isolator and associated equipment than micro-condensation. Thus, overall, the most desirable condition is probably that of sustained micro-condensation.

4.4 Validation of the Sporicidal Gassing Process

The gassing process has historically proved a major stumbling block in the validation of aseptic isolators, taking far longer than anticipated to complete. It is not entirely clear why this should be the case but lack of understanding and poor management may possibly contribute to the situation. The prime validation task will be cycle development and performance qualification (PQ). It is suggested that cycle development should form the main part of the gas generator operational qualification (OQ) and that the final confirming cycles form the PQ, but users may take differing views.

In virtually all cases, the goal of the cycle development will be *a complete cycle which delivers a reliable kill of the chosen biological indicator (BI) and returns the isolator to working condition, in the minimum time-frame.*

There are no definitive 'rights' and 'wrongs' in the validation process; the following account is just one view of the work. It is suggested that cycle development is made up of two parts:

- the establishment of 'worst-case' sites for BI tests
- the development of a cycle which gives the results described in italics above.

Challenge to the gassing process is usually by use of a number of BIs distributed through the isolator, bearing in mind that it is the surfaces of the isolator and not the air space that is to be treated. The sites for BI placement therefore need to be chosen logically. A good challenge primarily uses 'worst-case' sites, and well-distributed sites secondarily. All this work must be carried out with the isolator in its fully-loaded, and thus most challenging, configuration. Thus the first task is to establish the load pattern (or patterns) to be used in the isolator, to draw this pattern very clearly and to photograph it. This then provides the operators with the model to be used for all subsequent production and re-validation gassing cycles. Contact area must be minimised and so the isolator load should be suspended on stainless steel hooks where possible, or alternatively placed on stainless steel wire racks, with free gas circulation space between all the load items.

With the load pattern established, the first stage in the 'worst-case' search should be visualisation of the gas flow to determine areas of poor gas circulation. The gas is not visible and so smoke, or better still, water mist, can be used. The gas generator is connected up and operated without peroxide to produce an air flow pattern which parallels the gas flow. The hose of the water mist generator is then passed around inside the isolator to reveal the air flow pattern. Any areas of poor air distribution or re-circulation eddies can be noted as 'worst case' sites for BIs. It is a wise precaution to video the mist pattern for future reference and as part of the OQ documentation package.

The next stage is thermal and humidity mapping. Areas of higher surface temperature will have lower condensation and thus slower kill. Areas of lower humidity will also have lower condensation and thus slower kill. Small T/RH transmitters are available and these may be placed around the isolator and operated while the gas generator is run, loaded with water (ideally water for injection (WFI)) and not peroxide. Readout from the transmitters should reveal any warmer areas and any areas of low RH which should then be targeted as BI sites. Ideally, the RH graphs from the transmitters should show a classic build-up, plateau, and tail-off pattern, forming an analogue of the peroxide concentration in real use.

The final stage may be mapping with chemical indicators which change colour on exposure to VPHP. These are distributed around the isolator and the gas generator is run with peroxide on a 'guesstimated' cycle pattern. The places where the CIs are slowest to change colour are then noted as sites for BIs.

The data from visualisation and mapping can then be reviewed and 'worst-case' sites logically chosen for the BIs. As a very rough guide, something like 10 BIs per cubic metre might be a reasonable challenge to the gassing system. All extremities should be covered, as well as the 'worst-case sites' revealed by the studies. There is little point in festooning the isolator with large numbers of BIs. If the gas distribution is so poor that this is thought necessary, then the issue of distribution needs to be addressed. After all, if gas distribution were perfect, then a single BI would be adequate.

Careful consideration is needed when choosing a suitable BI for validation. Further consideration, combined with a written rationale, is needed to deal with the matter of 'rogue' BIs. These issues are fully evaluated in the PDA monograph on the subject of BIs for sporicidal gassing, in final draft form at the time of writing (Steele, 2006). Briefly, it has been recognised that BIs can give inconsistent results with previously well-characterised gassing cycles. Research shows that BI quality can vary from one manufacturer to another, apart from any inherent variability in the organisms. 'Rogue' BIs, ie those which grow during incubation, even after exposure to a well-developed gassing cycle, will almost inevitably appear from time-to-time. It is therefore reasonable to accept these, within suitable limits (Templeton and Hillebrand, 2005). The nature of the surface on which test spores are placed has been shown to affect the reaction to gassing cycles (Sigwarth and Moirandat, 2000) and validation engineers might consider using stainless steel, glass, plastic and Viton® carriers to prove the cycle valid for all isolator materials. However, in practice, a convention has emerged and the majority of validations take place with the challenge organism.

Spores of *Geobacillus stearothermophilus*. Spores of *G. subtilis* are also resistant to VPHP but *G. stearothermophilus* is generally more available since it is widely used in autoclave work. Some users go as far as sampling the wild microflora in the isolator to establish the real challenge.

- Loading at $1 - 9 \times 10^6$. Loading at 10^5 is probably perfectly acceptable and much easier to manufacture in reliable form, but is not the current convention. This situation could change in future.

- Stainless steel carrier. Stainless steel seems to be quite acceptable as an analogue for all the isolator surfaces.

- Tyvek® envelope. Some users have opted for inoculated carriers, as opposed to BIs with a Tyvek® envelope. This requires the carrier to be dropped into media tubes while in the isolator, but this saves the work of transferring and opening the envelope under LAF. The inoculated carrier is perhaps a better analogue of the natural micro-flora to be found on the isolator surface.

The BIs are usually attached to the walls of the isolator, equipment, sleeves, etc using small sections of masking tape or similar. It is suggested that one end be taped down with the gas-permeable face of the envelope away from the surface. The envelope may be bent away from the surface a little, so that it is exposed to the gas flow.

Once all of this preliminary work has been completed, the real work of cycle development can begin. There are several approaches to this process, for example:

- *The D-Value Method.* (Sigwarth and Moirandat, 2000) The first step in this method is to estimate the system D-value by exposing a bank of BIs, sited in the middle of the isolator, to a long gassing cycle and removing samples at regular intervals. The BIs are incubated and reported as 'growth' or 'no growth'. Limited Spearman Karber Method (LSKM) statistical analysis is then used to derive a D-value for the gassing process. The D-value method assumes that a plot of the remaining viable spore population on a BI (log10) against gassing time (linear) is a straight line. A gassing period of 10 times the D-value forms the validated cycle time, this giving a 4-log overkill on a spore population of 10^6. The method has been used quite extensively by the Swiss isolator manufacturer SKAN Ag (Figure 2.8).

- *The Kill Curve Method.* This method similarly exposes a bank of BIs to a long gassing cycle with sample removal at regular intervals, but the BIs are incubated and then enumerated. The results are plotted to give the system 'kill curve'. By extrapolation of the plotted curve, the time at which all the BIs have been killed can be estimated. This time is multiplied by between 1.5 and 2.0 to give the validated cycle time. The method has been used quite extensively by the UK company BioQuell Ltd (Figure 2.9).

- *The Kill Time Method.* This seeks to establish the 'kill time' for BIs in the whole isolator volume, without recourse to the D-value. In this case the full suite of BIs is set up in the loaded isolator, as developed from the worst-case studies. The isolator is then gassed using a cycle time derived essentially from the experience of the gas generator suppliers. This may appear naive, but the major suppliers have wide experience and can no doubt propose a realistic initial test cycle for most isolators. This initial cycle is termed the 100%

Figure 2.8 Straight Line Plot

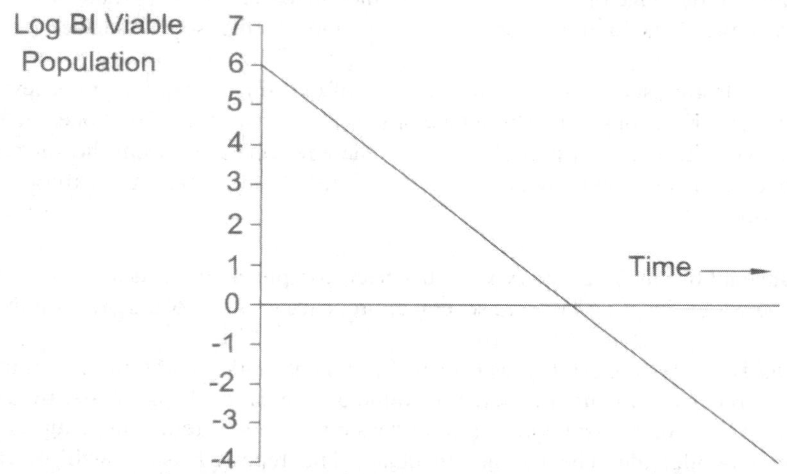

Figure 2.9 Kill Curve Plot

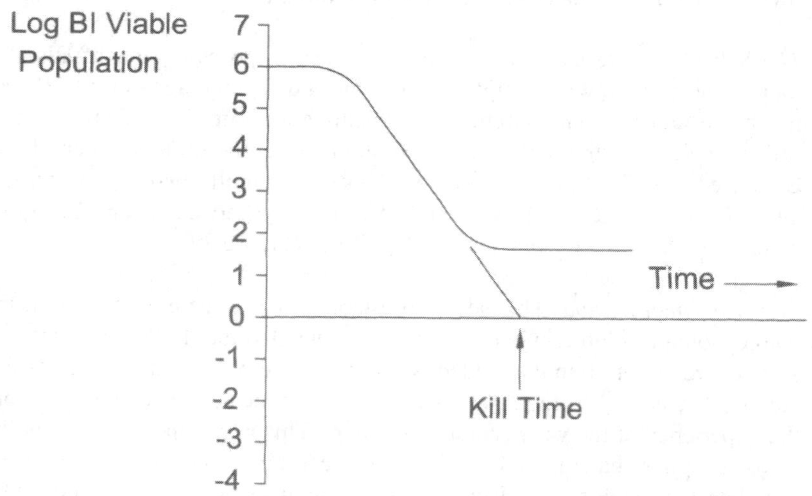

cycle. The BIs are incubated and reported for 'growth' or 'no growth'. If no growth occurs, the cycle time is halved (50% cycle) and the process repeated. If growth then occurs, the cycle is repeated at 75%. If no growth occurs at 75%, then the 'kill time' lies between 50% and 75% and it is reasonable to take the mean value of 62.5% of the original cycle as the kill time. This value is multiplied by 1.5 to give the validated cycle, the additional time applied as the kill time is not precisely established. If on the other hand, the 100% cycle shows growth, the cycle time is doubled, and so on. Thus it is possible to estimate the kill time with as little as three test cycles. The bold validation engineer may choose to run the 100%, 50% and 200% cycles in immediate succession to reduce the overall time requirement. In this way the gassing cycle may be developed quite rapidly. This method has not been used extensively but may find use in the future since it offers savings in validation time.

The final stage of the gassing process, the aeration of the isolator to remove the VPHP to a safe level, must also be validated. The 8-hour occupational exposure level (OEL) of VPHP is usually given as 1 ppm and so this is the normal target for aeration. This can be measured by the very simple Dräger tube method, or by instruments such as the Dräger Polytron or UOP 'Guided Wave'.

Some operators use an instrument to measure the concentration of the VPHP during the cycle, both in cycle development and subsequent production. As previously mentioned, such instruments exist and can perform this task. However the results may not be consistently accurate. Infra-red absorption instruments are more accurate but are not simple to use.

4.5 Ozone and Chlorine Dioxide

Ozone (O_3) and chlorine dioxide (ClO_2) have both been used as gaseous sporicidal agents. Ozone is very effective and breaks down to oxygen but is highly toxic and very corrosive, as it is a strong oxidising agent. Chlorine dioxide has a number of advantages and is marketed as a viable alternative to VPHP but requires an exhaust duct to atmosphere. Thus this open-loop system suffers an inherent disadvantage; however both of these agents may see new developments in future.

5 MONITORING

5.1 Physical Monitoring and Leak Testing

As with any other pharmaceutical process equipment, the continued function of isolators must be monitored and recorded to achieve compliance. The work divides roughly into physical and microbiological monitoring, with physical checks generally taking place at regular intervals and microbiological checks taking place more or less continuously. The main physical checks are usually carried out at six-monthly intervals and may be referred to as Planned Preventative Maintenance (PPM). A suitable protocol must be written and approved before PPM work starts.

5.1.1 Filter testing

The inlet and exhaust HEPA filters will normally be tested in situ, using smoke generated from an approved oil such as Shell 'Ondina EL' or Emery 3004 PAO, and a calibrated photometer. Testing should be carried out in accordance with ISO 14644-3. Current practice seems to favour the cold smoke generators (Laskin nozzle type), which need only a source of compressed air to operate. Hopefully the isolator will be equipped with a test port to introduce the smoke, and a further test port to check the smoke concentration immediately upstream of the filter.

The filters should be tested with the isolator in normal operating mode. For isolators with unidirectional down flow, the face of the inlet HEPA filter(s) and its seal must be scanned using the photometer head inside the operational isolator. It is not normally possible to scan the exhaust filter(s) and thus a volumetric measurement has to be carried out. The same is true of cartridge-type inlet and exhaust HEPA filters and ISO 14644-3 describes a test for such filters. Again, the isolator should be fitted with appropriate test ports to measure the smoke concentration downstream of the exhaust filters or cartridge filters.

Users may have their own site standards for HEPA filters but ISO 14644-3 calls for a maximum penetration of 0.01%. Some users require a maximum penetration of 0.001% for inlet filters and 0.01% for exhaust filters. Filter test engineers should take care to close off all test ports after the work has been completed.

5.1.2 Instrument calibration and alarms

The operational instruments of the isolator need to be calibrated. The pressure meter should be calibrated from zero to full-scale deflection on both rising and falling regimes. A well-designed isolator will have tapping points on the front face to allow the test instrument and air pump to be connected easily and quickly. If the isolator has HEPA filter DP gauges, these too must be calibrated. Calibration of air flow may not be entirely simple. In the case of unidirectional down flow isolators, the air velocity off the filter face can be used. For turbulent flow isolators, a pitot tube or a hot-wire anemometer can be used, ideally at a point in the ducting where the flow is reasonably straight. Alternatively the exhaust (or inlet) duct might be fitted with a temporary adaptor so that a vane anemometer can be used directly. The ideal flow calibration instrument is an orifice plate, set in a long tube to give straight flow. This is then connected to the section of duct carrying the isolator flow transducer and an external variable speed fan is used to blow air through the system. An external fan is not needed during maintenance if the isolator has circuitry which allows the engineer to vary air flow over the full available range.

While the pressure and flow instruments are taken through their range during calibration, we have an ideal opportunity to check the alarms. It should be possible to pinpoint the setting at which alarms are triggered, both high and low excursion. Any other instruments such as humidity or oxygen level can normally be calibrated according to the maker's instructions.

5.1.3 Set pressure hold

Most isolators are designed to maintain a set canopy pressure (the pressure with respect to the isolator room) within the practical limits imposed by entry and exit from sleeves and half-suits. This function should be checked at PPM by repeated disturbances while monitoring the pressure.

5.1.4 Particle counting and recovery

Particle counting results are to some extent, the 'proof of the pudding' for isolators, as they are for cleanrooms. If the HEPA filters are sound and the isolator is leak-tight, then good particle-burden conditions will result. The concept of 'at rest' and 'in use' does not really apply to isolators and so they are usually counted with the ventilation in normal operation but with no operators present. The number of counting sites is an issue — unless the isolator is large, the figure of the square root of the floor area in square metres will be impractically low. A common-sense approach suggests that a two-glove isolator might need two particle counting sites, a single half-suit isolator might need three sites and a large filling line isolator five or six sites. The next dilemma stems from the current version of Annex 1 (EU GMP) with its poorly-conceived logic for sampling volumes. Two scenarios have emerged for isolators. In the first, a full cubic metre of air is taken at each sampling site. In the second, a total of one cubic metre of air is taken from the total number of sampling sites.

If possible, the particle counter should be placed inside the isolator so that sampled air is returned to the isolator. If this is not possible, the isokinetic head sample tube may be taped though a glove which has a finger tip removed. In this case, the isolator pressure may affect the counter air flow rate. ISO 14644-1, ISO 14644-2 and ISO 14644-3 provide the various classifications of air cleanliness and thus after particle counting, the compliance of the isolator can be established.

In some cases, it will be necessary to check the recovery rate of the isolator and a typical challenge measures the time taken to purge the particle burden down from ISO Class 7 conditions to ISO Class 5 conditions. A surprisingly small quantity of smoke is needed for this test and it is suggested that a party balloon be partially inflated using the smoke generator, then sealed. The balloon is transferred into the isolator and burst using the gloves, to raise the particle burden appropriately (Herriott, 2005).

5.1.5 Breach velocity

Containment isolators are operated at negative pressure so that any leakage will be inwards. In the event of a major leak, containment isolators need to maintain negative pressure and also to maintain a minimum inflow velocity of 0.7 m/s through the aperture. This is deemed sufficient to protect the operators, at least in the short term, from exposure to the contents of the isolator.

This feature should be tested at PPM and the standard practice is to remove a glove to give an aperture about 100 mm diameter. The inflow velocity is then measured with a vane or hot-wire anemometer at the centre of the cuff. In some cases a sleeve is removed and the velocity at the shoulder ring is taken.

Very high inflow velocities should be avoided and while there is no statuary upper limit, anything above 3.0 m/s is suspicious. The reason for this is that eddies and back-flow may develop at high air speeds and actually move material out of the open aperture.

5.1.6 Leak Testing

Leak testing for isolators is quite a major issue and indeed commands an entire chapter in the 'Yellow Guide' (Midcalf et al, 2004). Perhaps the first question to raise is 'why do isolators need leak testing anyway?'. Positive pressure isolators will keep contaminants out while negative pressure isolators will keep materials in, despite minor leakage. However, leaving aside the case of aseptic containment isolators (eg cytotoxics), the overall integrity of the isolator logically forms a part of its compliance remit. ISO 14644-7 and ISO 10648-2 establish four classes of leak-tightness or 'arimosis' which can be used to characterise isolators. It is up to the isolator manufacturers and users to specify which of the classes is appropriate for a given application.

It is important to note at this stage that two quite separate processes are involved here: leak detection and leak rate measurement. The two should not be confused.

Table 2.1 Isolator Arimosis Classes

Class of isolator ISO 14644-7	Hourly leak rate (hr^{-1})	% Volume loss per hr	Volumetric leak rate m^3s^{-1} (for a $1 m^3$ isolator	Standard decay time (mins)	Single hole equivalent (microns diameter)
1	$<5 \times 10^{-4}$	<0.05	0.14×10^{-6}	>1.5	103
2	$<2.5 \times 10^{-3}$	<0.25	0.70×10^{-6}	>6	232
3	$<1 \times 10^{-2}$	<1.0	2.8×10^{-6}	>30	464
4	<1 × 10–1	<10	N/A	N/A	N/A

Leak detection is used to establish the presence of leaks and locate the site of leakage. Various methods are available. Smoke testing employs the same equipment as HEPA filter testing and the smoke particles are of similar dimensions to microorganisms, forming a good representative challenge. The isolator is filled with smoke, sealed and held at pressure, while the photometer sampling head is moved along the various seams, seals and joints of the isolator structure. Negative pressure isolators are often tested at positive pressure but it is possible to put the photometer head inside a negatively-pressurised isolator and waft smoke along the seams and seals outside. The smoke method is quite sensitive but leaves the isolator with a coating of oil, which needs to be removed.

Helium gas can be used to detect leaks with a suitable electronic detector or 'sniffer'. The isolator may be sealed and then taken to a suitable test pressure (3–5 times working pressure) using helium gas. The helium detector is then applied to the isolator structure to show the site of leaks. This method is clean and inexpensive, is frequently used, but is relatively unsatisfactory. Helium is very mobile and the site of apparent leaks often cannot be reproduced at a second pass of the detector.

Ammonia gas is widely used by French isolator manufacturers for leak detection. Special bromophenol-treated cloth is available, which turns from pale yellow to bright blue in the presence of ammonia gas. A petri dish of 33% ammonia solution is placed in the isolator, which is then sealed and brought to test pressure. The cloth is applied to the isolator structure steadily and leaks are quickly revealed by the colour change. The method is sensitive and inexpensive and, though labour-intensive, is very useful.

In contrast to the process of leak detection, leak rate measurement is used to quantify the leakage, thus to place the isolator in the required ISO Class and ensure compliance. Again various methods are available and are well-described in the 'Yellow Guide' (Midcalf et al, 2004). However, in practice, most operators will use the pressure decay method. This simply involves sealing the isolator, raising it to a suitable test pressure (again, 3–5 times working pressure) and monitoring the rate at which the pressure decreases. The method is quite sensitive, easy and practical to perform and requires little test equipment, other than a calibrated micromanometer and an air pump. In some cases the isolator may be 'self-leak testing', using its own fan and pressure gauge.

The snag with pressure decay testing is that it is subject to interference from several sources. The first of these originates with the isolator sleeves and/or half-suits, which are by their nature flexible and liable to change their volume during the test. The universal gas law applies here and thus if the volume of the isolator changes by 1%, the pressure will change by 1,000 Pa. It is this effect which generally renders automatic leak tests relatively valueless.

Where half-suits are involved, they may be removed and the aperture capped, but it is probably better to leave the suits in place and test them as part of the isolator structure. In this case the suits must be rigidly suspended and not hung from elastic cord, which will simply allow the suit to move up and down with the isolator pressure. The best plan for sleeves is to evert them fully (ie pull them out of the isolator) during the pressure decay test. In this condition, it is arguable that they act as pressure-compensators but at least consistent results can be obtained. The same is also true for the canopy of flexible film isolators, which are to some limited extent elastic.

Negative pressure isolators tend to be more difficult to pressure test, or rather to obtain consistent results. Many users resort to testing negative pressure isolators at positive pressure.

Further sources of interference come from changes in atmospheric pressure and isolator internal temperature which may take place during the test. Atmospheric pressure can occasionally change by as much as 10 mb (1,000 Pa) per hour and generally changes by as much as 1 mb (100 Pa) per hour. The significance of atmospheric pressure change is less with high test pressure and short test times but even tests to Class 3 could be affected. If the isolator internal temperature changes by 1.0 degree C, the gas law shows a resulting pressure change of about 350 Pa. Again, this could have an effect on the test result and thus the test engineer and the isolator users should develop a clear policy in relation to these effects. They may, for instance, decide to monitor atmospheric pressure and external (ie room) temperature and accept the pressure decay test results provided that no major changes have taken place during the test.

Tests to Class 1 however, cannot be considered valid without measurement of atmospheric pressure and internal temperature, and correction of the initial results accordingly. The extra test equipment consists of a barometer resolving to 1 Pa (0.01 mb) and a thermometer resolving to 0.01°C. The isolator pressure reading at the end of the decay test is corrected thus:

- if the atmospheric pressure has gone up, add 1 Pa to the final pressure reading for every 0.01 mb and vice versa

- if the isolator internal temperature has gone up, subtract 3.5 Pa from the final pressure reading for every 0.01°C, and vice versa.

Apart from the physical problems associated with leak rate measurement, the user is faced with the question of what test pressure and what time of decay should be used to establish the arimosis class of his isolator. No actual figures are given in any of the standards or guidelines and the following process is therefore proposed.

- The test pressure may be between two and five times the working pressure. Lower test pressure does not present sufficient challenge and higher test pressure may damage rigid isolators. A reasonable and supportable figure is three times working pressure, thus around 150 Pa in many cases.

- The test time may be between five and 30 minutes. Less time does not allow the isolator to 'settle' and more time gains little information and is more prone to atmospheric and temperature changes. A reasonable and supportable figure is 15 minutes.

- Having decided on the test pressure and the test time, the results are applied to the following equation:

$$L = \frac{PD \times 6000}{SP \times M}$$

Where

L = the leak rate in % volume loss per hour
PD = pressure decay in Pa
SP = 101,325 + the starting pressure in Pa
M = the test time in minutes (Watling and Parks, 2004)

The figure for pressure decay (PD) may of course, be corrected for atmospheric pressure and internal temperature effects.

Thus, armed with a clear and supportable figure for the leak rate of his isolator conforming to the appropriate arimosis class, the user can comfortably claim full compliance.

5.2 Microbiological Monitoring

In some ways isolators can be considered as small cleanrooms and monitored as such. Thus settle plates form a useful microbiological monitoring system, with perhaps four plates placed around the periphery of a 4-glove isolator. These may be exposed in the non-operational condition or during operation, but exposure time must not be excessive or the plates will dry out. Active air sampling is often used to back up settle plate data in isolators. In some cases a self-contained sampling machine is used, while some isolators have the facility for a built-in sampling head, with the air pump outside the isolator.

Contact plates or swabs are used to check the microbiological burden of the isolator surfaces, under a developed standard operating procedure (SOP). Care should be taken to clean the site of contact plate application to remove residual nutrient agar. Swabs have the advantage of reaching less accessible areas, particularly on isolated equipment such as filling machines. Finger dabs may be taken at the end of a work session to confirm the aseptic status of the gloves. Again, residual nutrient needs to be wiped off afterwards.

Table 2.2 Microbiological Monitoring Type and Frequency

Frequency	Activity
Each work session	Finger dabs Settle plates in the isolator during operation
Weekly	Settle plates in the isolator out of operation Surface sampling of the isolator Settle plates in the isolator room Surface sampling of the isolator room
Three-monthly	Active air sampling in the isolator and in the isolator room Media fill (if applicable)

In the aseptic isolator, there should be no recovery of viable microorganisms from the environmental monitoring process.

6 VALIDATION FOR ISOLATORS

6.1 Introduction

In the past, the validation of isolators, especially major installations such as filling lines, appears to have taken a disproportionate length of time and effort. The reason for this is not clear but may have been caused by a lack of experience on the part of both the users and the regulatory authorities. Isolators have now been in full-scale industrial use for around 20 years and so this should not still be the case. The validation of isolators may therefore be approached in the same way as any other piece of pharmaceutical process equipment, with a clear plan and a set of approved protocols.

A suggested overall validation structure (which could be applied to most pharmaceutical equipment) is laid out in Figure 2.10. It should be emphasised that whatever validation structure is applied, the documentation must follow a clear, logical and sequential identification system. Such a system will allow an inspector to examine any stage of the validation process with an overall view of the documentation covering not only that stage, but those which precede and follow it. It is suggested that the eventual validation flow diagram may be annotated to illustrate the document identification system used. Methods such as these will clarify the process for all those involved, produce quick approvals and speed the validation process, thus ensuring compliance.

It should be noted here that validation is not an item to be tacked on at the end of an isolator project. It has to be considered at the outset and integrated throughout the life of the project continuing on into production. Furthermore, validation costs money, and this fact must be recognised in the project budget. As a rough guide, validation should account for around 10% of the overall project cost.

6.2 Isolator Validation Protocols

The Validation Master Plan (VMP) is the high-level document which gives an overview of the whole project, of which the isolator may form either a small or large part. Naturally, the isolator validation work must conform to the strictures of the master plan.

Figure 2.10 Validation Flow Diagram

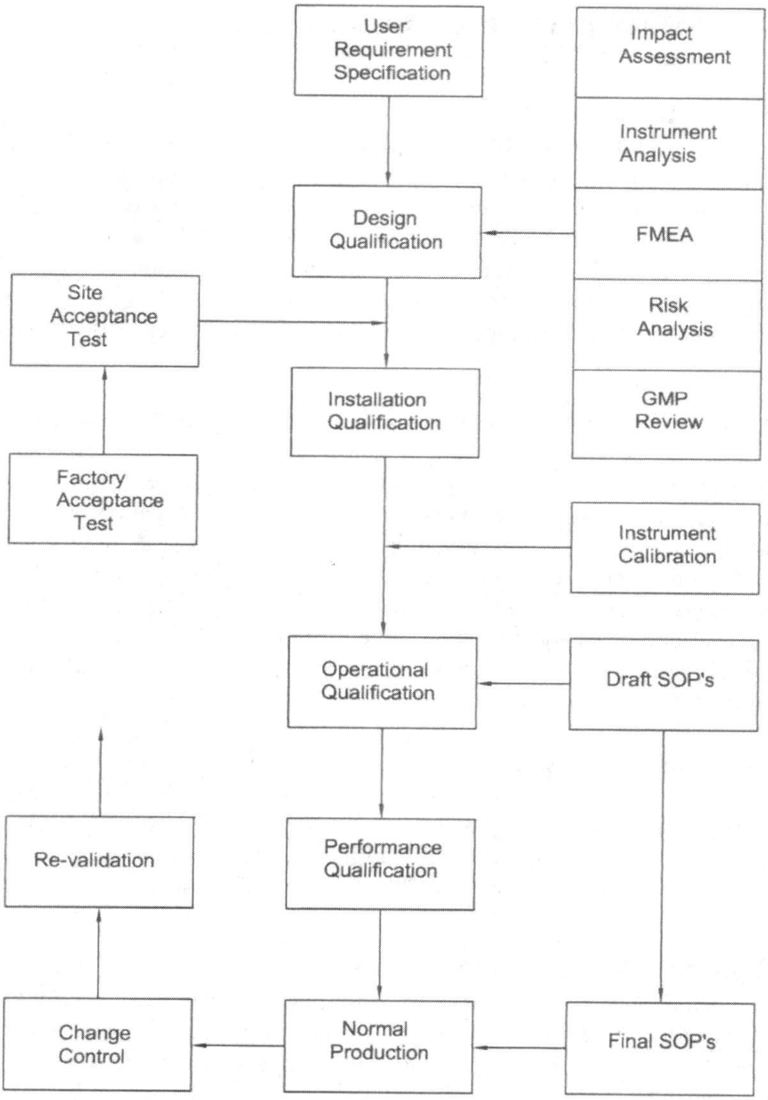

The User Requirement Specification (URS) is written by the users to describe what they want the isolator to do. This can be done by laying out a series of specific requirements. At the same time, however, it should have a degree of flexibility to allow the potential suppliers some freedom of design. It can be an advantage to write the URS in co-operation with a chosen supplier. As an example, an isolator URS might be conveniently divided into the following sections:

- the process to be carried out
- the process equipment to be accommodated in the isolator
- the process equipment to be interfaced with the isolator
- materials to be introduced
- products and wastes to be removed
- utilities required
- the requirements for sterility or containment
- standards for the isolator environment (eg ISO Class 5)
- construction and finish
- control, instrumentation and alarms
- room or building limitations (eg floor plan, door widths, etc)
- standards and guidelines to be applied
- ergonomic and safety issues.

Note that specific design details are avoided in this list. For example, ISO Class 5 may be the URS requirement, but it is left to the isolator designer as to how it might be achieved. In this case, the designer would probably respond by fitting unidirectional downflow ventilation.

The URS is a very fundamental document in terms of compliance because it is against this document that the entire performance of the equipment will be judged. Thus time spent developing an accurately-worded URS is time well invested for the future of the project.

The Functional Design Specification (FDS) is normally produced by the suppliers to describe in detail how their design will go about conforming to the URS. It should include not only verbal descriptions but also drawings and specifications.

The Design Qualification (DQ) is not so much a protocol but a review process, which may be documented through minutes of a formal DQ group meeting. Thus the users and suppliers may convene to review the FDS, perhaps line by line, and thus assess its conformity to the URS and also to other standards or guidelines such as GMP and risk analysis.

Following DQ, the suppliers can go into manufacture, leading to the next phase of validation, which is the Factory Acceptance Test (FAT). An approved protocol is needed here, written to check the general performance of the isolator in terms of satisfactory construction and finish, process ergonomics, pressure and flow operation, and instruments and alarms.

The isolator can then be shipped to site and installed ready for the Site Acceptance Test (SAT). This too, should have a written and approved protocol, which will probably be more detailed than the FAT. Indeed the SAT may embrace many of the checks which could be carried out under Installation Qualification (IQ) and OQ. In this case, to avoid repetition of work, the IQ and OQ may largely take the form of a review of the SAT. Thus the SAT protocol would need to test conformity to the URS and FDS.

If sporicidal gassing is involved, then cycle development may be required and this can be documented under the OQ protocol. Performance Qualification (PQ) is usually carried out by the users and may consist of in-process quality checks such as air quality and microbial contamination. The PQ may also include three sporicidal gassing cycles with BI's to confirm the cycle developed under the OQ.

After each of the 'Q' stages (IQ, OQ, PQ) have been executed, it is now conventional to write a brief report, which summarises their findings. Each report is filed ahead of its respective protocol, in the assembled project validation file, thus an inspector can review a qualification stage quickly using the report, or in depth using the full protocol.

The now-classical validation process is potentially repetitive and a relatively new concept has developed to circumvent this issue. Integrated Commissioning and Qualification (ICQ) attempts to avoid duplicating work carried out at FAT, only to be repeated at SAT and perhaps even again at OQ. In essence, this involves more extensive FAT documentation which is then used to support more limited

OQ documentation. This view of validation seems likely to develop further in future in order to simplify, clarify, and above all shorten the process of moving pharmaceutical plant into revenue-earning production.

6.3 On-Going Validation

If the isolator is to continue to produce good results, it must be maintained throughout its operational life, a process which might be termed 'on-going validation' or 'planned preventative maintenance' (PPM). Once again, an approved protocol should be adopted for this work, which may very well incorporate elements of the OQ. Thus a typical six-monthly isolator PPM protocol might specify the following checks:

- general condition assessment
- general function and alarm check
- pressure hold check
- instrument calibration
- HEPA filter testing
- leak testing
- particle counting.

As with all the 'Q' protocols, each of these checks should be laid out in the general format:

- title
- purpose
- scope
- test equipment
- criterion of acceptance
- table of results
- pass/fail statement

- comments
- attachments (eg test instrument calibration certificates, instrument readouts, etc).

Should any of the validation protocols need to be altered or updated at any time, then this must be carried out under a formal change control procedure and correctly documented.

7 RESTRICTED ACCESS BARRIER SYSTEMS (RABS)

7.1 Introduction

The historical difficulties encountered with sporicidal gassing and validation have to some extent set in train the development of a hybrid containment system, now known as the RABS. This device is structurally mid-way between the open cleanroom and the totally sealed isolator. It takes various forms with greater or lesser degrees of exchange with the room air, while the more sealed forms may still be subject to sporicidal gassing, despite the apparent problems. The RABS is, of course, used for aseptic operations rather than toxic containment.

Perhaps the typical RABS has unidirectional downflow regime and looks very much like an isolator down as far as working height. At this point the air is exhausted to the room through a series of openings or grills, rather than passing through exhaust HEPA filters. Thus the RABS containment relies more heavily on engineered air flow and less heavily on physical barriers. It is suggested that the RABS is easier and quicker to validate than an isolator, but retains most of the benefits. It is not entirely clear why this should be the case, especially if the RABS still uses sporicidal gassing. It is also suggested that the capital cost of a RABS is less than that of an isolator and that equipment alterations (eg vials size change) are easier. The RABS should perhaps be considered as an upgraded cleanroom, rather than a downgraded isolator.

7.2 Definition

In 2005 the ISPE set up a committee under the chairmanship of Richard Friedman of the FDA to examine the RABS, to produce a definition and some guidelines. The committee produced a brief but useful three-page document in August 2005 (Hillebrand and Templeton, 2005). The design and operation paragraphs are quoted here in full by courtesy of the ISPE Committee.

1 of 3 **Restricted Access Barrier Systems (RABS) for Aseptic Processing** *ISPE Definition August 16, 2005*

Introduction Human operators pose the greatest risk to product contamination during "conventional cleanroom" aseptic processing. Many different barriers of varying capabilities have been used to separate operators from critical sites during aseptic processing with the objective of reducing the probability of a contaminated unit. These range from simple flexible curtains used on many traditional aseptic processing lines to *advanced* aseptic processing in isolators.[1] A Restricted Access Barrier System (RABS) is an advanced aseptic processing system that can be utilized in many applications in a fill-finish area. RABS provides an enclosed environment to reduce the risk of contamination to product, containers, closures, and product contact surfaces compared to the risks associated with conventional cleanroom operations. RABS can operate as "doors closed" for processing with very low risk of contamination similar to isolators, or permit rare "open door interventions" provided appropriate measures are taken.

Design A RABS provides a level of separation between operator and product that affords product protection superior to traditional systems. There is no single design model for a RABS; however these systems share the following common "quality by design" characteristics:

- Rigid wall enclosure[2] that provides full physical separation of the aseptic processing operations from operators.

- Unidirectional airflow systems providing an ISO 5 environment[3] to the critical area.

- Sterilization-in-place (SIP) is preferred for contact parts such as fluid pathways. Where this cannot be achieved, such parts should be sterilized in an autoclave, transferred to the RABS via a suitable procedure and aseptically assembled before processing. Product contact parts such as stopper feed and placement systems should be sterilized in an autoclave and aseptically assembled before processing.

- Entry of material such as environmental monitoring materials, consumables, containers and closures shall be via a suitable transfer system that prevents exposure of sterile surfaces to less clean classification environments.

- Use of glove port(s), half suit(s) and/or automation to access all areas of the enclosure which need to be reached by an operator during filling operations.

    — Gloves and gauntlets attached to glove ports are required to be sterile when installed; thereafter, gloves should be sanitized or changed as appropriate to minimize the risk of contamination.[4]

- "High-level disinfection"[5] of all non product contact surfaces within the RABS with an appropriate sporicidal agent before batch manufacture.[6]

- Surrounding room classification should be ISO 7 minimum in operation.[3]

2 of 3

- Some processes may include rare open door interventions. In these cases, because of the inherently increased risk to product, the following are required to maintain the RABS protection concept:

  — Provision for disinfection of non product contact surfaces using appropriate decontaminating agents following a door open intervention.

  — Locked door access or interlocked door access with recorded intervention alarms (and/or other satisfactory means of documentation) and mandated appropriate line clearance.

  — Positive airflow from the enclosure to the exterior environment while the door is opened.

  — Appropriate ISO 5 classification[3] areas may be necessary immediately adjacent to outside of enclosure to always assure ISO 5 conditions[3] inside the RABS. Examples of such situations are:

    — Setup of oversized sterile equipment that requires unwrapping of autoclave packaging outside of the RABS.

    — Any machine sections that require open door interventions (such as certain powder filling applications). After any open door intervention, operations cannot re-start until ISO 5 conditions[3] are re-established in the critical zone.

Operation When aseptic filling operations are carried out in a RABS, it is the intent to eliminate the need to open RABS doors after thorough sporicidal disinfection of non product contact surfaces. Design to prevent door openings can be achieved by a number of measures which include Clean-In-Place/Sterilize-In-Place (CIP/SIP) to the point of fill for liquid filling operations, remote or automated sampling for in-process control testing (IPC) including monitoring for viable and non-viable particles, and the use of enclosed transfer systems which offer greater protection during introduction of components and pre-sterilized equipment. Another key measure includes automation, use of glove ports, and/or use of half suits to eliminate need for "open door" interventions. When open door interventions are necessary[7], an ISO 5 (class 100)[3] vertical unidirectional airflow system outside of the RABS reduces risk of a breach in ISO 5 conditions[3] and further safeguards the aseptic integrity of the system. Each intervention that requires opening of a door of the RABS is regarded and documented as a significant event. Interlocked RABS doors facilitate control and documentation. Following an open door intervention, appropriate line clearance and disinfection commensurate with the nature of the intervention are required as outlined in the standard operating procedures.

1 A decontaminated unit, supplied with ISO 5 (100 part/ft^3 at 0.5 microns) or higher air quality, which provides uncompromised, continuous isolation of its interior from the external environment (e.g., surrounding cleanroom air and personnel). *Guidance for Industry — Sterile Drug Products Produced by Aseptic Processing — Current Good Manufacturing Practice*, September 2004.

ISO 14644 Clean rooms and associated controlled environments — Part 7: Separative Devices (clean air hoods, gloveboxes, isolators, and mini environments), 2004.

EUDRALEX Volume 4 — Medicinal Products for Human and Veterinary Use: Good Manufacturing Practice — Annex 1, May 2003.

2 There are occasions where containment of toxic materials is required and special closed or containment RABS may be used.

3 All room classifications are "in operation" and include appropriate microbial action levels in table 1 of *Guidance for Industry — Sterile Drug Products Produced by Aseptic Processing — Current Good Manufacturing Practice*, September 2004.

4 Section V, Personnel Training, Qualification, and Monitoring, *Guidance for Industry — Sterile Drug Products Produced by Aseptic Processing — Current Good Manufacturing Practice*, September 2004.

5 High-level disinfectants are defined as "capable of destroying all organisms with the exception of high numbers of resistant spores", adapted from Block S, *Disinfection, Sterilization, and Preservation*, Fourth Edition, 1991, and the *Manual of Clinical Microbiology*, Sixth Edition, p.232.

6 In certain circumstances, multiple day operations are possible depending on design, appropriate disinfection plan, risk mitigation steps, early regulatory review (i.e., pre-operational review is recommended), and a subsequent ongoing evaluation of process control data.

7 Product contact parts need to be sterilized and their sterility must be maintained. When appropriate, fitting of oversize equipment (such as stopper feeding parts) during filling setup should be followed by appropriate and thorough disinfection (product contact parts need to be sterilized).

An important point (Lysfjord, 2004) is that the RABS is just that — a system and thus it can only function provided that all of the constraints including the quality of the room, operator gowning, training and the like, are observed.

7.3 Active and Passive RABS

RABS have been divided into two broad groups — 'active' and 'passive'. In the active RABS, open-door access is allowed during processing, under certain defined and controlled conditions.

In the passive RABS, no such access is permitted during processing and so any process equipment must be totally reliable.

7.4 Conclusion

The RABS appears to offer the product quality level of a true isolator but at reduced cost, both initial and running costs and for these reasons we may see further development and application of the technology.

It is expected that the Pharmaceutical and Healthcare Sciences Society (UK) (formerly the Parenteral Society) will produce a monograph on the subject of RABS in the near future.

8 ISOLATOR OPERATION

8.1 Siting

In the early days of isolator technology, it was thought that since isolators were sealed, they could be operated in almost any type of room conditions. However, practical considerations apply, for instance transfer processes place a challenge on the isolator system, the challenge being greater if the room conditions are poor. At the same time, even enclosed pharmaceutical processes must be carried out under some defined, controlled conditions. Thus it became clear that isolator rooms would need to be built to certain general standards. There were early attempts to develop a matrix table which proposed room conditions appropriate for specific processes (eg TPN Compounding) when using specific types of transfer port. However, these seem to have fallen by the wayside in favour of blanket guidelines.

Before the advent of ISO 14644, since EU GMP Grade D was the lowest grade of controlled conditions defined, this quickly became the minimum condition for aseptic isolator operation. Grade D is roughly between ISO Classes 7 and 8, thus many users will opt to build their aseptic isolator rooms to achieve ISO Class 7 conditions at rest. The more cautious may go to Class 6 — justified if the isolator uses open transfer devices such as the mousehole. Occasionally, Class 5 rooms are specified outside the isolator, perhaps where other critical processes are also taking place.

To meet these classes, the isolator room will effectively become a cleanroom and its construction will therefore reflect this — materials must be non-shedding, crevice-free, easily cleaned and resistant to normal cleaning agents. In particular, equipment such as hand wash basins must not be present in the room. The isolator room should be entered via a changing room of the same ISO Class. If there is more than one room in the isolator suite, the standard pressure cascade regime should be applied with 15 Pa differential between each room.

In hospital pharmacies the isolator room is commonly sited next to a support room, which supplies materials for processing and accepts the resulting products back. Ideally such a room should maintain the same conditions as the isolator room. Transfer between the two rooms will often be via interlocked pass-through hatches, preferably having one dedicated to inward transfer and another dedicated to outward transfer.

If the isolators are used for toxic or pathogenic containment, then the rooms will of course, be run with a negative pressure cascade. In either case, the users may wish to establish the 'cleanup time' for the isolator room, that is, the time taken to go from poor conditions at start-up to operational condition. This can be checked with a particle counter and may be done at the same time as the general particle monitoring tests.

The 'Yellow Guide' (Midcalf et al, 2004) reproduces a useful table prepared by the NHS and the MHRA (UK) listing isolator room conditions against various application, sanitisation methods and transfer devices. While aimed at hospital pharmacy uses, the table gives useful guidance for industry operations, though it does only refer to GMP room classification, rather than ISO classification.

8.2 Clothing

The standard of cleanroom clothing will be dictated by the ISO Class at which the system is operated, thus full cleanroom gowning will be needed for ISO Class 5 while much less rigorous regimes can be adopted for ISO Class 8.

Some users specify only a lab coat, overshoes and mob cap to be worn in the Class 8 isolator room. Others go a stage further and include a Tyvek® suit, which considerably reduces the particle burden shed by the operator.

Whichever clothing regime is chosen, the changing room should incorporate a step-over system, with hand-washing facilities on the 'dirty' side. Jewellery such as wristwatches and sharp rings must be removed, not only for general hygiene reasons but also to prevent damage to isolator sleeves and half-suits.

Gloves are an issue in isolator rooms, since the operator will be using a further pair of gloves to access the isolator. It seems reasonable to wear light latex gloves in the isolator room provided that these fit into the isolator gloves without problems. This 'double-gloving' acts as a further barrier between the operator and the isolator contents, perhaps giving an added degree of security where, for instance, cytotoxic materials are being handled.

8.3 Training and Operation

It is important to train those who work with isolators, not only in the details of the isolator but in good working practices generally. Serious incidents have taken place (Farwell, 1994) as a result of poor practice, stemming from lack of operator training. It has to be clearly understood that isolators are not magic boxes that guarantee sterility and that they must be used with appropriate GMP procedures at all times.

Thus users should be generally aware of the principles of isolator design and function and fully aware of the controls and instruments fitted to their isolators. They should carry out the various daily, weekly and monthly checks before using the isolator and monitor certain instruments during processing. In particular, operators need to know what to do in the event of any given alarm or alert condition, especially in toxic or pathogenic applications.

As with any pharmaceutical process equipment, each operator should maintain an up-to-date training record.

8.4 Cleaning

Regular cleaning is an absolute requirement for isolators whether in aseptic or toxic use. In both cases, this will reduce the challenge to the isolator as a safety barrier between operator and process, thus minimising the chances of contamination or a containment failure and maximising the degree of compliance. Section 4 of this chapter (pp66–75) has discussed the process of sporicidal gassing, where it was noted that cleaning is an integral part of the bio-decontamination process and thus an essential pre-requisite to gassing. So basic is the cleaning process that it should be carried out under an SOP, by trained personnel and appropriately recorded. There are broadly two stages to the cleaning process.

- Gross cleaning can be summed up as 'sweeping up the broken glass and mopping up the spills'. This means dealing with the most obvious, visible and large-scale build-up of undesirable materials, using whatever tools and methods are most appropriate from dustpan and brush to lint-free wipes. Used cleaning materials have to be removed safely and disposed of correctly and of course, this particularly so where toxic or pathogenic debris is concerned. Disposable RTP containers (see Section 3, p60) are singularly useful in this case, reducing the exposure of the cleaners to an absolute minimum. It clearly pays dividends to consider cleaning for isolators at the design stage.

- Detail cleaning is designed to remove the surface debris that may not be easily visible but is none the less potentially harmful to the isolated process. In some cases it may also form a disinfection process — normally consisting of a wiping down exercise, using lint-free wipers moistened with a cleaning or disinfecting agent. The wiping action should take the form of parallel, overlapping strokes which reach all the isolator surfaces. It may be a good plan to use one of the proprietary extended reach cleaning tools to achieve this successfully.

A wide variety of cleaning agents is available and some suppliers are very knowledgeable in this area, so their advice can be freely sought. For aseptic applications it is desirable to use a sporicidal agent. Where toxic materials are involved, there may be a particular cleaning agent which de-activates the toxin, eg dilute sodium hydroxide solution renders many cytotoxic agents relatively safe.

The chosen cleaning agent has to be compatible with the isolator materials. For example, sodium hypochlorite is often used to treat pathogenic isolators but the chloride ions present will attack even the highest grades of stainless steel. In this particular example, hypochlorite can be used provided it is fully removed after the required contact time. Alcohol (70% IMS or IPA) is often used as an inexpensive and effective cleaning agent — but it is not sporicidal and is flammable. Alcohol can also affect the canopy of flexible film isolators, but only if used frequently and in large volumes, which leave the plastic wetted with alcohol for long periods of time.

Some specialised process isolators may be fitted with CIP for which suitable solvents and cycles will need to be validated. Indeed, the complete cleaning process may need to be validated using physical tracers such as riboflavin or paracetamol and/or microbiological tracers such as swabs or contact plates, depending on the application of the isolator.

8.5 Maintenance

As mentioned in Section 6.3 (p87), isolators require maintenance. PPM has been discussed but there are a number of minor maintenance tasks which need to be carried out between the relatively infrequent PPM checks. These will normally be described by the isolator manufacturer but might take the form of the following.

Daily checks:

- canopy pressure correct
- air flow rate correct
- HEPA filter DP gauges within limits
- gloves undamaged
- sleeves undamaged
- half-suits undamaged.

It is worth noting at this point that research has shown visual inspection of gloves by trained operators to be more effective at detecting leaks in gloves than any other test method (Sigwarth and Moirandat, 2000).

Weekly checks. All the above plus:

- re-filter elements clean/clear
- door/port seals undamaged
- flexible film canopies undamaged.

Monthly checks. All the above plus:

- alarms operational
- breach velocity correct
- windows sealed and secure
- support framework secure and undistorted.

All maintenance work should be logged by the operators at the time it is carried out. Such logs should be shown to the PPM engineers so they are aware of any problems or issues with the isolator.

REFERENCES

Farwell, J. (1994) Aseptic Dispensing for NHS Patients (The Farwell Report) *Journal of Hospital Infection* **27**: 263–273.

Herriott, N. (2005) HEPA Services Ltd. Personal communication.

Hillebrand, J. and Templeton, P. (2005) *Rogue Biological Indicators and Re-validation*. ISPE Barrier Isolator Technology Forum. Prague, September 2005.

HSE/MHRA (2003) Handling Cytotoxic Drugs in Isolators in NHS Pharmacies.

Lee, G. and Midcalf, B. (eds) (1994) *Isolators for Pharmaceutical Applications*, HMSO.

Lysfjord, Jack (2004) Global expert on sterile manufacturing. Personal communication.

Midcalf, B., Philips, M., Neiger, J. and Coles, T. (eds) (2004) *Isolators for Pharmaceutical Applications*. Pharmaceutical Press.

Sigwarth, V. (2005) *Gloves — Evaluating Leak Tests*. Management Forum Conference. London. January, 2005.

Sigwarth V and Moirandat P. (2000) Development and Quantification of Decontamination Cycles. *PDA Journal of Pharmaceutical Science and Technology*, August/September.

Steele, G. (Committee Chairman) (2006) *Biological Indicators for Sporicidal Gassing*. PDA draft monograph.

Thomas, P.H. and Fenton-May, V. (1987) Protection Offered by Various Gloves to Carmustine Exposure. *Pharmaceutical Journal* **238**:775–777.

Watling, D., Parks, M. (2004) The relationship between saturated hydrogen peroxide, water vapour and temperature. *Pharmaceutical Technology Europe* 16, March 2004.

3

CAVEATS OF BACTERIAL ENDOTOXIN TESTING

Kevin Williams

A 'caveat' has been described as both a warning or caution, and a qualification or explanation. The word is used here with both meanings to help form a list of issues to have in mind when devising and performing tests for bacterial endotoxins. In its broadest sense the goal of such testing is to preclude the occurrence of significant levels of pyrogens from drugs, drug constituents, and drug containers/closures. Any oversight that impedes or lessens the likelihood of such preclusion or misjudges the levels that can be considered as 'significant' can be viewed as deserving of a caveat. Furthermore, any historical context or elaboration of test mechanism that broadens our views or deepens our understanding of the test deserves mention as well. This is one user's list of top caveats — at best, pet peeves, or pet projects at least*.

USING THE IN-PLATE SPIKE METHOD

The advent of spiking samples into 96 well plates† brought greater simplicity to the kinetic test and reduced testing by removing a parallel series of tubes. However, spiking in the plate can be misapplied where the sample treatment is harsh or otherwise depyrogenating. It is not the goal of endotoxin testing to

---

* For discussion of these and other issues visit *http://tech.groups.yahoo.com/LALUserGroup/*

† Personal communication with Robert Blumenthal, Cambrex, Walkersville, MD.

Figure 3.1 In-Plate Spike Method

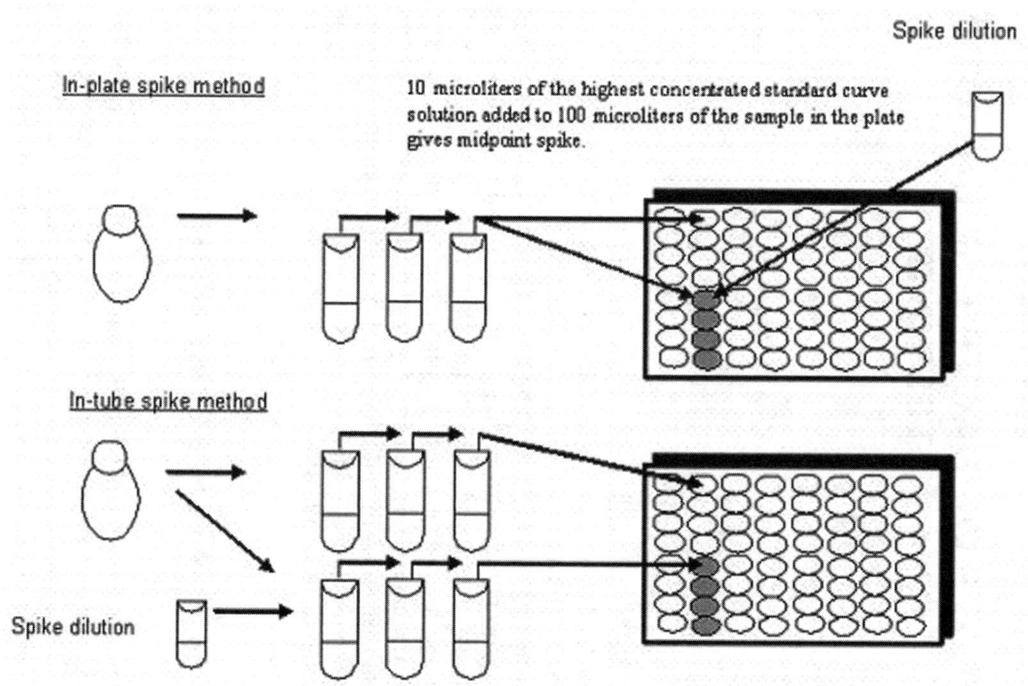

Reconstitution in harsh solvents, extreme pH, adsorption, etc. Anything that could decreased or inactivate the recovery of endogenous endotoxin in the sample dilution should be spiked in a parallel tube series to endure the entire sample treatment.

destroy potential endotoxin in a sample prior to testing, but such could be the case if one blindly applies the in-plate spiking method. The initial treatment could very well reduce or completely eliminate the endotoxin content of a sample prior to serial dilution and testing, particularly with poorly water insoluble compounds, or for compounds that must be suspended in acid or base.

For example, if one reconstitutes a drug product in buffer, adjusts the pH to highly acidic levels to aid dissolution and then subsequently serial dilutes in buffer or water, then the positive product control recoveries may well be near perfect, but would reveal nothing about the depyrogenation that has occurred in the initial sample dilution. This is a fine point but a critical error that could go unnoticed until a pyrogenic episode revealed it. Figure 3.1 illustrates the concept of tube versus plate spike. The demonstration of the recovery of spike from the entire reconstitution and subsequent dilution process should be performed during validation, so that for routine testing one may employ the spike method. If the process is found to diminish endotoxin recovery then a less harsh method should be developed and validated.

DOSING/SPECIFICATION DEVELOPMENT/SAFETY FACTORS

The initial validation process may be as ongoing as the final drug product itself, with many factors subject to change during drug development — the product potency (PP), presentation, included excipients, interference factors, containers, etc. Factors absolutely critical to establishing a test that will detect the endotoxin limit concentration include the maximum human dose (MHD), product potency, LAL lambda (λ) to be used in the Tolerance Limit (TL) and Maximum Valid Dilution (MVD) or Minimum Valid Concentration (MVC) calculations. An error in calculation or failure to secure a relevant dose for the TL calculation will nullify subsequent efforts to provide an accurate result. The TL is equal to the threshold pyrogenic response (K in EU/kg) divided by the dose in the units by which it is administered (mL, Units, or mg) per 70 kg person per hour. Factors to watch in this critical calculation include:

(a) Adjusting for the body weight (conversion from m^2 may be necessary as for many cancer drugs).

(b) Clarifying the means of delivery (bolus versus multiple daily doses, etc) as endotoxin is cleared from the body so the relevant dose is that which is delivered per hour.

Figure 3.2 Overview of BET Laboratory Compliance and Control

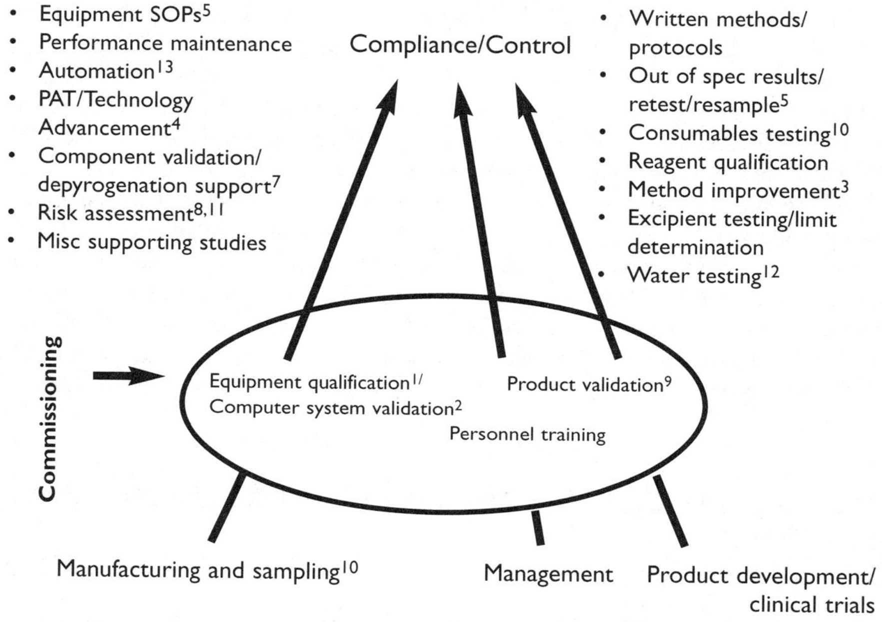

Selected References

1 Huber L. (2002) Validation of computerized analytical and networked systems. Interpharm Press, Englewood, CO.
2 21 CFR Part 11. http://www.fda.gov/cder/guidance/5667fnl.htm
3 Cooper, JF. (1990) Resolving LAL Test Interferences. J Science & Technology 44(1), Jan/Feb, pp.13–15.
4 http://www.fda.gov/cder/guidance/6419fnl.htm/
5 United States District Court for the District of New Jersey, USA, Plaintiff v Barr Laboratories Inc., Civil Action No 92–1744.
6 USP–29 NF–24 Chapter <85>.
7 21 CFR 211:80 Subpart E — Control of Components and Drug Product Containers and Closures.
8 http://www.fda.gov/cder/gmp/index.htm.
9 FDA Guideline on Validation of the Limulus Amebocyte Lysate Test as an End-product Test for Human and Animal Parenteral Drugs. Biological Products and Medical Devices, Dec 1987: http://www.fda.gov/cber/gdlns/lal.pdf
10 http://www.fda.gov/cder/guidance/5882fnl.htm.
11 http://www.fda.gov/oc/guidance/gmp.html.
12 USP–29 NF–24 Chapter <1231>.
13 USP–29 NF–24 Chapter <16>.

(c) Basing the dose on a method of delivery relevant to the means of administration, usually on the units of active ingredient (ie do not use mL instead of mg).

(d) Adjusting the MVD formula calculation for a potency change. The formulas exist as a set so that the potency as a drug is reconstituted at (say 10 mg/mL versus 20 mg/mL) has to be consistently applied in protocols for testing from validation to routine testing. One cannot base the calculations on a 10 mg/mL reconstitution and later decide to use a 20 mg/mL reconstitution.

(e) Increasing the dose in the clinic to a level that exceeds that used as a basis for MVD calculation in the testing laboratory. Typically, in clinical testing, the dose of a given drug is varied, sometimes widely. The dose used should be the maximum, and the greater the communication as the study progresses, the better. One does not want the testing laboratory to be limited by a large dose that is not going to be used as it makes the test more stringent than necessary. Neither does one want to assume that the doses given in the clinic are smaller than they actually are, as this would not preclude the occurrence of pyrogenic episodes (the reason for testing in the first place).

While it is not the responsibility of the assay development laboratory to set specifications, the personnel assigned to this lab do play a key role in verifying that the specifications set are within the appropriate bounds established by the FDA Guideline calculations and pharmacopoeial requirements. Practically speaking, the laboratory will determine the informal specification for early development testing given the clinician's dose range. At a later date project leaders will discuss the specifications with the company's specification or standards committee with an eye toward including some additional safety factor. There appear to be divergent philosophies on setting specifications. The first is to set the most stringent specification the laboratory can support (ie around the limit of detection). The second is to set the specification around the regulatory limit allowed (ie the tolerance limit calculated value), which is the highest legal level. Table 3.1 illustrates the effect of dose changes on the resulting tolerance limit (the scientific basis of forming specifications). An additional consideration is the capacity and associated variability of a manufacturing process to remove endotoxin which will be dependant upon its method of manufacture.

Setting the specification too tightly may come back to haunt the participants in the form of a test failure and subsequent destruction of an expensive lot of drug that scientifically, and from a regulatory perspective, does not exceed allowable endotoxin levels. Early clinical doses are often several-fold higher than subsequent marketed drug doses, but there does not appear to be a mechanism to ratchet specifications down as doses decrease in the clinic. When products inevitably go to market, they will do so with a dose that is sometimes significantly lower than that used to establish the endotoxin test. The second philosophy is as poor as the first. If the specifications are set too close to the values allowed by law, then the routine examination of the drugs will not detect changes in endotoxin content until they are at failing levels. Ideally, one wants to 'see' the endotoxin content well below the specification to serve as a warning that the manufacturing process is beginning to allow contamination well before it reaches a relevant level. If the specification is too high, there will be no time for corrective action preceding a test failure.

Table 3.1 Effect of dose changes on the resulting tolerance limit

	Development compound dose	Clinical dose	Marketed dose
	2,000 mg	1,000 mg	250 mg
TL*	0.175 EU/mg	0.35 EU/mg	1.4 EU/mg

*TL = K/M

A reasoned solution for setting balanced specifications is the seemingly common task of assigning a 'safety factor' to specific articles depending upon their place in the manufacturing process. For example, for drug product (so called end-product) the calculated tolerance limit is the legal limit and often one feels like an additional safety factor, even a fractional one or a smaller rounded number is typically appropriate. Some have sought to assign such a safety factor in a manner that is consistent. For example, for active pharmaceutical ingredient (API), Dr. Cooper has recommended assigning a 'safety factor' of $^1/_4$ the assigned end product limit (Williams, 2004). Similarly methods to assign limits have been developed by Williams and Cooper (Williams, 2001) based on either the proportionate amount of a given excipient in the drug formula, or based upon the amount commonly present in 100 mL of a given small volume injection as a worst case. These two methods will be discussed later in this chapter.

STARTING A LAB

The overall process is important in the development of a new LAL assay for a drug to be used in clinical studies. If the process has too many 'one offs' then it may be necessary to double check the many pieces of the overall puzzle. Establishing a process that captures all the details is critical to ensuring that the right tasks are performed in the right sequence, the right information is documented, and this is then correctly applied to the test, both in its performance and in the determination of the parameters that govern its performance. Such a detailed process may be difficult to capture in a standard operating procedure (SOP) and extensive training will be necessary before an analyst is proficient in all the nuances of developing an LAL assay, particularly for a new drug candidate. There are three broad categories necessary to start a new lab (see Figure 3.2):

- equipment qualification including Computer System Validation (CSV)
- people (training and testing practices)
- product testing (including validation, specification formation, etc).

The GMP documentation expectation for any analytical test is that of being able to 'recreate' the test including all the materials used in a given assay. For the LAL assay that can be a daunting task if the right systems are not in place. In looking back on any test there may be dozens of consumables and equipment references (water or other diluent preparation, LAL, CSE, tips, tubes, plates, pipettes, tips, containers, water bath or heating block or kinetic reader, or other equipment, analyst, etc) for which lot numbers must be recorded and according to the USP, a demonstration made that the articles do not interfere with the test.

The checklist of activities is exhaustive and forms a useful overview relevant to starting a lab from scratch. For instrumentation and equipment preventative maintenance and calibration records are needed — for personnel, somewhat analogous to instrumentation, qualification (training) and documentation of the events.

For product validation documentation, certificates of analysis or other proof that the consumables used do not inhibit or enhance the test, RSE/CSE and/or COA reagent qualification documents used are all part of the items needed to 'back up' any given test. Printed laboratory notebooks or worksheets should be used to collect all the pertinent information in an organized fashion.

Communication with new chemical entity submitters (compound owners) and clinical end users (those dosing the clinical drugs) is also important. With regard to developing tests for unique chemical entities, a decentralized lab will have the advantage of compound knowledge but may lack sufficient depth of assay expertise. Conversely, a centralized lab may have depth of assay knowledge but limited compound and compound manufacturing process knowledge. Ideally, the test lab will have access to both knowledge sets.

APPRECIATING THE SENSITIVITY OF THE TEST/ AVOIDING A SKEWED VIEW OF THE ASSOCIATED ERROR

The original two-fold associated 'error' of the LAL test still referenced in the USP was associated with the gel clot assay in which a series of two-folds was used to determine the actual sample content. As a result one could not 'see' in between the serial dilutions so the true result always lay between the two tubes. In today's kinetic testing the results are extrapolated between tubes (between standards) on the linear line formed by the curve. With the gel clot assay one could test the same sample a hundred times and never know where between those two tubes the 'true' result lay. However, with kinetic testing one could test the same sample many times and the results would not be off by as much as 50–200% one from another. Thus the true variability associated with a kinetic test may vary from lab to lab. However, it would be surprising, if with experienced analysts and qualified pipettes and reader, the error were greater than 15–20% in the same lab.

The associated 50–200% 'error' of the test can create the false impression that the associated sensitivity of the test is poor. However, with lambda as low as 0.005 EU/mL* or fractions of a nanogram, the test is more sensitive than any chemical test for endotoxin. As an analogy, consider a 50–200% error in the fitting of a pair of shoes or pant — this would seem unacceptable indeed. Compare that to the same error associated with threading a needle which certainly seems exacting enough. The latter is analogous to the error associated with the detection of endotoxin (thread) with an LAL (needle). Determining minute levels of endotoxin to 50% or even 200% can be a difference of less than a nanogram, an infinitesimally small physical amount of endotoxin.

Consider in this discussion the sensitivity, or lack thereof (insensitivity) of the early rabbit Pyrogen assay. The sensitivity of the assay can be gauged as the 5.0 EU/kg rabbit threshold pyrogenic response multiplied by a 4kg rabbit weight divided by a 10 mL dose of drug product, which comes out to about 2.0 EU/mL. Compare this with LAL assay sensitivities that go as low as 0.001 EU/mL. This suggests an approximate range of up to 2000-fold greater sensitivity for the LAL test (Figure 3.3).

---

* Even 0.001 EU/mL for ACC Pyros Kinetix tube.

Figure 3.3 Detecting the endotoxin 'thread' via the LAL 'needle'. At great sensitivity an error of 50–200% is small.

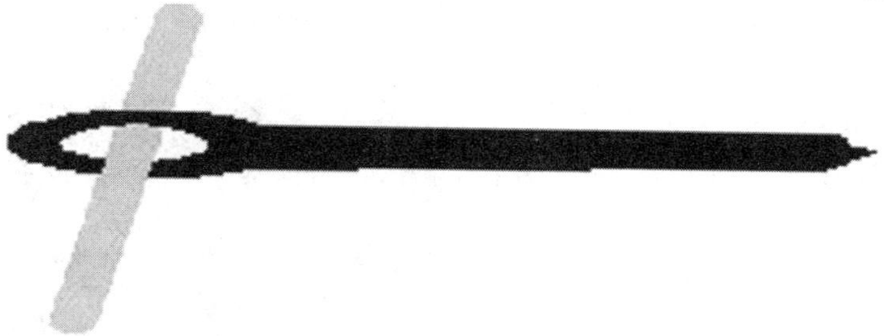

CONTAINER CLOSURE TESTING/DIFFERENTIATING DEPYROGENATION DESTRUCTION AND REMOVAL

Historically, medical devices have been tested and regulated in a different manner to drug products. Their non-invasiveness has fostered appropriate leniency in testing, including testing composite samples (average results resulting from multiple samples combined in a single rinse). However the CFR (CFR, 2004) makes it clear that the expectation is that closure/containers are different items than medical devices.

The worry with container/closures is that any washed item that sits has the capacity to hold water in the little pools inherent in their shape (ie stopper wells). This water can then grow gram-negative organisms — notorious for their negligible growth requirements and ability to proliferate in water. The subsequent evaporation of the water will leave the bacterial residue in the washer wells (ie endotoxin). Thus the treatment of these items is expected via a three-log reduction process.

A caveat in the testing of container/closures is the inherent difference in treating such items by the two processes. Washing and baking are by far the predominant methods of use in the industry. The two methods differ in the number and control over variables. For instance, the application of heat is well known and can be uniformly delivered and measured by physical parameters. However, washing includes variables such as the solubility of contaminants (ie endotoxin),

Figure 3.4 Disc Seals with Clinging Bubbles

mixing, heat, etc. These parameters are empirical and thus cannot be predicted but must be demonstrated by testing.

Component testing via washing and rinsing contains some predictable caveats. These include:

- Developing an optimum process for removal is necessary. For example, bubbles that occur during washing cling to stoppers or disc seals (shown as example below in Figure 3.4). Surfaces create areas that do not interact with the wash process, and thus remove applied or endogenous endotoxin poorly.
- The area to which concentrated endotoxin should be applied is the component surface that contacts the product (the relevant area for 3-log reduction).

Figure 3.5 Applied endotoxin too easily removed says nothing about the process undergoing validation

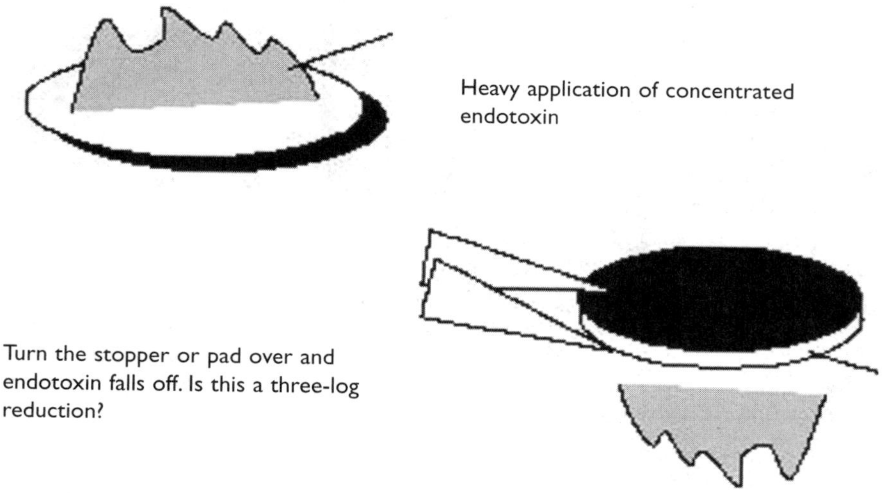

- The amount of 3-log reduction achieved depends upon the amount of endotoxin applied — the greater the amounts applied, the greater the ease of their removal. One can envision that if too much is added, let us say 1,000,000 EU/component, it will be too easily removed and will reveal little about the depyrogenation method employed (see Figure 3.5).

- Solubility properties are important in washing processes and do not enter into baking processes as significant variables. See the 'baby and the bathwater' analogy (Figure 3.6). A theoretical problem associated with cleaning validation studies relating aptly to depyrogenation validation (endotoxin removal) studies using a 'tar baby' analogy has been described by Agalloco (Agalloco, 1992):

  > The cleanliness of the bath water may not necessarily relate directly to the cleanliness of the baby. If the contamination is not soluble in the cleaning agent, then the contamination will remain on the surface. If the contamination is not soluble in the final rinse, samples of the bath water will not detect the presence of residual contamination. The conclusion will be drawn that the baby is clean, when in fact both the cleaning and evaluation methods are inadequate.

Figure 3.6 Baby in the bathwater analogy

giving 'tar baby' a bath

baby's bath water

tests 'clean' data infer that the baby is clean as the water is clean but baby is still dirty

In other words, if one determines the cleanliness of the stopper (baby) by measuring the endotoxin ('tar') remaining in the bath water (laboratory rinse method), then one has to ensure that the method used does indeed remove the 'tar'. There must be some validation of the method to serve as a demonstration it removes endotoxin from 'sticky' surfaces. At least theoretically, endotoxin that clings tenaciously to a stopper (thereby escaping detection and allowing a poor removal process) can be removed later by the surfactant action of a drug, and become available for parenteral administration.

DEVELOPING AN ENDOTOXIN CONTROL STRATEGY

There are various ways to arrive at test specifications for various drug constituents including API, bulk, excipient and raw materials. However, arguably the 'best' way is to view all the constituents of a given drug product in relationship with one another as they occur in the final product. Thus, each can be viewed in proportion to their relative contribution with regard to additive risks. Many have developed complex methods of gauging risks (likelihood of occurrence times the severity of the risk for example). However, for BET, the risks are well known to be related to the dosing of the drug material. Thus, each constituent of the drug product —

water, containers, raw materials, closures, and API — are coming together in a constant stream with known contributions.

The rivers that flow toward the final destination include:

- glass and rubber (negligible contributors as they are clean to a level of three logs below any activity)
- water (which has a USP associated limit of nmt 0.25 EU/mL)
- raw materials
- excipients (some of which have compendial limits but most of which do not)
- API — which one can use Cooper's rule of $^1/_4$ the TL of the end-product drug.

For the raws and excipients that do not have compendial limits one can use one of two methods previously described by Williams (Williams, 1998) and Cooper (Cooper, 2004) respectively. The former involves assigning limits to API and excipients present in the drug formulation in proportion to their presence and thereby assigning a limit relevant to their participation in the final drug dose. This avoids an arbitrary assignment when compendial limits are not available, as they often are not. A formula for the method is given below:

TL (drug substance with excipients)

$$= 350 - \frac{((TL_{e1} \times W_{e1}) + (TL_{e2} \times W_{e2})...)}{WA}$$

where TL_{e1} is the tolerance limit of excipient 1 and W_{e1} is the weight of excipient 1 per dose of active drug and WA is the weight or units of active drug per dose. Note that the formula ((...)) indicates all relevant excipients without an exclusion rationale should be included in the calculation. Compare the result obtained to the end-product tolerance limit calculated in the formula: TL = K/M for a final 'sanity check'.

Cooper's method takes advantage of a common feature of small volume parenterals (SVP) in that they are limited to doses of 100 mL:

> The diversity in the use of excipients makes it a challenge to devise a uniform strategy for selection of limits and test protocols. One could simply set an arbitrary limit or

assign limits based on their proportion in an SVP formulation, as proposed by Williams. However, excipients have one common attribute to exploit. An SVP is limited to 100 mL; this volume can represent the dose for calculating an endotoxin limit. A compendium of excipients was published that details the range of concentrations for excipients in SVP formulations (10). A uniform way for calculating an excipient endotoxin limit (EL) is proposed that is dependent on the maximum amount of excipient in 100 mL of an SVP:

$$\text{Excipient EL} = \frac{\text{350 EU (adult endotoxin tolerance limit)}}{\text{Maximum amount of excipient in 100 mL}}$$

$$= \frac{\text{3.5 EU/mL}}{\text{units/mL}}$$

The latter method seems best for developing broad, cross-applicable specifications for various excipients perhaps on a corporate or compendial level (and Cooper includes such a table in the reference), while the former method is suitable for specific drug products, as a 'sanity check' of API/excipient specifications, and as a necessity when uncommon or large amounts of excipients are employed.

HAVING AWARENESS OF VARIOUS BACTERIAL ENDOTOXIN TESTING RISKS

To aid discussion let us separate out some overlapping BET risks to include:

- regulatory risk
- sampling risk
- activity risk.

Regulatory risk

The concept of adulteration in drug and device manufacture stems from the Federal Food, Drug, and Cosmetic Act, which considers that a drug or device is adulterated if it does not comply with the provisions outlined in the Act (Munson, 1993). A drug is considered adulterated if it is 'filthy, putrid or decomposed in whole or in part' (Sharp, 1995), if its strength differs from that claimed, or if impurities are present. Though endotoxin content is not required to be absolutely absent, as are organisms in sterility testing, the definition of 'adulteration' cannot leave room for legal disagreement. Ultimately, the method listed in the USP monograph for a specific method serves as the final arbitrator of contamination from a legal standpoint. If the drug is listed in the 'official compendium' (USP in the US) and the strength, purity or quality is below the specific requirements listed in the monograph for that item, then it is considered adulterated by the FDA. According to Avallone (Avallone, 1986), the FDA does recognize different levels of risk associated with different types of violations. He cites the situation of a sterility test failure as a very serious health hazard worthy of a Class I recall classification. The citation of a more general lack of sterility assurance (ie observations, but not a direct analytical failure) due to the lack of cGMP conformance is considered a Class II recall classification (a less serious health hazard). Avallone gives three levels of philosophical compliance relating to the retesting and resolution of sterility failures exemplified by manufacturers (Sharp, 1995; Madsen, 1994):

(i) ... those who recognize the many limitations of aseptic processing and of the sterility test (when an initial positive test result occurs, they reject the batch)

(ii) ... others have this same concern when only gram-negative microorganisms show up on the initial test ... due to the virulence of the organism and because the organism would be more indicative of a process problem than one which might occur during testing (ie lab contamination)

(iii) ... other manufacturers ... have still another level of GMP philosophy and will do everything to justify the release of a product failing an initial sterility test.

It is not FDA's responsibility to develop and promulgate the best methods for each specific type of analytical testing required to demonstrate the safety, identity, strength, purity, and quality of drugs. The USP serves as a repository for this purpose to a large extent (in the US) and is an interface between government regulators and industry. The USP is cautious in the unique role that Congress has allowed it and change as a result of consensus often occurs slowly (Rothschild, 1990). Manufacturers have felt, historically from time to time, that guidance can be lacking. However, from a regulatory perspective, it is perhaps best (from the vantage of protecting public safety) to make manufacturers aware of their general obligations and even to let them worry if they are meeting them, but not to dictate the best scientific solution for each situation. It is in these areas that a given manufacturer's scientific and validation proficiency will determine compliance success. In such cases the critical science that the regulations were formed to ensure cannot be neglected. Does the method make sense? Does it really prove what it purports to prove? Has a worst-case validation philosophy been applied? What can go wrong in the test? What is the worst that can happen? The manufacturer should be the expert in the manufacture of pharmaceuticals and contamination control with the appropriate regulatory body overseeing that the valid requirements that have accumulated over time (ie the current in cGMP requirements) are applied appropriately to ensure that the public will receive safe and effective medicines. In short, a sample does not have to 'fail' the BET to be subject to regulatory risk of noncompliance.

Sampling Risk

In any microbiological test there is some chance that enough sample will not be taken to reveal the true amount of contamination that may be present in an article. The sterility test is a notorious example of this. The BET requires sampling from beginning, middle, and end of the lot and the 1987 FDA Guideline on Validation of the LAL Test (FDA, 1987) takes this concept further by stating that 'The sampling technique selected and the number of units to be tested should be based on the manufacturing procedures and the batch size. A minimum of three units, representing the beginning, middle, and end, should be tested from a lot. These units can be run individually or pooled.' The assumption is that contaminants are homogeneous throughout a given manufactured lot but the Guideline's call for basing the sampling upon the number of units from different parts of the lot makes it apparent that there must be some associated risk of a non-homogeneous contamination with endotoxin.

Often end-product testing is thought to be the absolute determining factor in producing a quality product free of contaminants. However, recent emphasis has been placed on in-process testing and process validation. These have become as integral as end-product analytical testing to providing the documented evidence that a product meets predetermined criteria of quality. The FDA Guideline on Validation of the LAL test as an 'end-product' test serves to illustrate the contrast between overtly stated requirements and current GMP (cGMP) expectations. The 'end-product' Guideline (FDA, 1987) makes no specific mention of testing requirements for items other than the end products, but clearly a manufacturer would be cited in any FDA inspection for failure to conform to industry expectations for ensuring the suitability of a given manufactured product at each major step of a manufacturing process. (CFR reference 312.23 includes IND requirements for drug substances even at the investigational stage.) LAL testing is routinely performed on the API, bulk products, excipients (particularly those of natural origin), water, cell culture media and additives, as well as vial and closure components used to contain the end-product.

Given the statistically low probability of finding a single contaminated vial in a given lot of product (at least for a non-homogeneous contamination mechanism), the USP chapter on 'Sterilization and Sterility Assurance of Compendial Articles' <Chapter 1211> describes the importance of process control in manufacturing. The following is presented in the context of sterility testing but can be applied as readily to endotoxin sampling and subsequent testing:

> ...this absolute definition (of sterility) cannot currently be applied to an entire lot of finished compendial articles because of limitations in testing. Absolute sterility cannot be practically demonstrated without complete destruction of every finished article. The sterility of a lot purported to be sterile is therefore defined in probabilistic terms, where the likelihood of a contaminated unit or article is acceptably remote. Such a state of sterility assurance can be established only through the use of adequate sterilization cycles and subsequent aseptic processing, if any, under appropriate current good manufacturing practice, and not by reliance solely on sterility testing (USP, 2000).

Though sterility is an absolute concept (ie contains no viable organisms), endotoxin content is not. There are acceptable, albeit low, levels of bacterial endotoxin allowed in drug products. The FDA (appendix E) and USP (monograph) publication of tolerance limits and provision of TL formulae make this distinction. Nevertheless, it does not change the dependence on probabilistic product sampling and, hence, the endotoxin test result, like the sterility test result, is not an absolute assurance of the lack of contamination except for the vial(s) consumed in the test:

> It should be recognised that the referee sterility test might not detect microbial contamination if present in only a small percentage of the finished articles in the lot because the specified number of units to be taken imposes a significant statistical limitation on the utility of the test results. This inherent limitation, however, has to be accepted since current knowledge offers no nondestructive alternatives for ascertaining the microbiological quality of every finished article in the lot, and it is not a feasible option to increase the number of specimens significantly (USP, 2000).

The regulatory precautions set in place are, in many (if not most) cases, due to the poor probability associated with finding spot contamination by quality control sampling techniques. The generally accepted sterility acceptance level (SAL) is 10^{-6} (ie one possible survivor in a million units), but according to Akers and Agalloco this value was selected as a convenience (Cooper et al, 2002). They maintain that 10^{-6} is a minimal sterilization expectation and should be linked 'to a specific bioburden model and/or particular biological indicator ... (otherwise) it is a meaningless number that imparts little knowledge on the actual sterilization process'. If the concept of sterility assurance seems somewhat less than rigorously defined, once — by necessity — one strays from the absolute definition, then the concept of the bacterial endotoxin tolerance limit can also be viewed scientifically (not legally from a compliance point of view) as existing with a range with variability of its own. Values making up the TL calculation include

- the use of an average patient's weight
- a maximum dose (that may be an actual dose subject to volumetric variances)
- a threshold pyrogenic response which has a range of its own
- the presence of endogenous endotoxin around the tolerance limt that is not *E. coli* LPS and is measured by a control standard endotoxin referenced against a reference standard endotoxin defined by an average potency that varies sometimes widely as determined in a variety of different laboratories.

In this context then the passing result obtained just under the tolerance limit can be viewed a single sampling point of an endotoxin content that may in reality be either above or below the TL cut-off. From a legal (compliance) standpoint, however, a single test may constitute a passing or failing (adulterated) product. Here again is a good argument for controlling the endotoxin content of products to levels well below that which creates a dilemma as to whether the product is contaminated at the TL.

Bruch (Bruch, 1993) relates that the PSI (probability of a survivor per item) for a can of chicken soup is 10^{-11} whereas the assurance provided by the USP Sterility test alone is not much better than 10^{-2}, given a twenty item sampling and is, as Bruch says, due to the rigorous heating cycles developed by the canning industry to prevent the possibility of survival of *Clostridium botulinum*. Bruch maintains that the industry has 'never relied on a USP-type finished product sterility test to assess the quality of its canned goods ... (because) the statistics of detecting survivors are so poor that the public confidence ... would be severely compromised through outbreaks of botulism'. He cites the generally accepted sterility assurance for a large volume parenteral item as 10^{-9} and 10^{-4} for a small volume parenteral that has been aseptically filled and sterile filtered as opposed to terminally sterilised. The apparent contradiction in the necessity of more stringent sterility assurance for a can of soup than for a parenteral drug is due to the ability of organisms to grow in soup as opposed to the likelihood of such growth in the parenteral manufacturing environment (see Table 3.2).

Table 3.2 Estimates of Probability of Survivors for Sterilised Items (from Bruch, 1993)

Item	Probability of survivor/unit
Canned chicken soup (a)	10^{-11}
Large volume parenteral fluid	10^{-9}
Intravenous catheter and deliver set(a)	10^{-6}
Syringe and needle(a)	10^{-6}
Urinary catheters(a)	10^{-3}
Surgical drape kit(a)	10^{-3}
Small volume parenteral drug (sterile fill)	10^{-3}
Laparoscopic instrument (liquid chemical sterilants)(b)	10^{-2}

(a) dosimetric release: no sterility test
(b) limits of USP sterility test 10–1.3 (with 95% confidence)

Activity Risk

Detection of endotoxin near the limit can be seen as an 'activity' risk, for want of a better term. There is some risk in disagreement of the exact level of the content, because any given result near the limit is based upon gauging the true value of an unknown level of true contamination that is near the legally allowed level. If the given product is tested more than once, as in a retest for example, then one can have conflicting determinations (albeit all generally close in value). To address the risk that such an absolute value determination (ie specification) is difficult to exactly determine, some so-called 'safety factors' are desirable to avoid quibbling if levels of endotoxin detected are close to the legal limit. (See discussion on 'safety factors'.) One does not want to box oneself into a situation of releasing a lot of drug product that is realistically at the threshold pyrogenic response level but that barely 'passes' based on limited sampling and test results. The best way to ensure that a product meets the requirements of the compendia is by using the best test one can develop at levels below the tolerance limit at various stages throughout the manufacturing process.

ENDOTOXIN AGGREGATION ISSUES (IE 'STICKINESS')/PROTEASES AND PROTEINS AS ASSAY DEVELOPMENT TOOLS

The purpose of developing a Bacterial Endotoxin Test is to overcome LAL-test interferences that exist between the article to be tested and the reagents used in the test, namely LAL and CSE (control standard endotoxin). A difficulty almost always inherent in the test is not interference *per se*, but rather, as Cooper says, 'an adverse effect of the test specimen or container on the aggregation state of purified endotoxin controls. That is, inhibition is often a failure to recover inadequately dispersed LPS, a state that may not represent the detection of environmental endotoxin contamination under the same conditions' (Cooper, 1990). Nevertheless, it is a condition that must be dealt with constantly as it is the inherent nature of purified lipopolysaccharide used as the control standard. This inherent 'stickiness' of endotoxin to itself and to hydrophobic surroundings is overcome by the addition of mechanical (vortexing and sonicating) and chemical (surfactants) energy by the user. There are two separate forces toward aggregation:

- the propensity of lipid A to form high molecular weight aggregates based upon its hydrophobicity
- the electrostatic charge (cationic) associated with the sugar component of the molecule whereby an increase in ionic strength of a solution brings about aggregation and thus poor recovery.

The first force is the basis for the necessity of extreme vortexing always associated with the preparation of the CSE. The latter force can be overcome by dilution or by the addition of cationic dispersants such as Pyrosperse® (Cambrex) or Tween (etc). Thus aggregation can occur via hydrophobic interaction with the lipid A end of LPS, or via hydrophilic association with the polysaccharide portion of the LPS molecule. These forces put the LPS molecule at odds with itself and ready to align rather with other molecules, like for like.

Related to this is the difficulty of testing some protein-containing compounds that have been found to 'mask' endotoxin. Petsch et al. (Petsch et al, 1998) describe an interference mechanism involving the properties of certain cationic proteins. They refer to this as 'masking' — an interference mechanism that appears somewhat rare compared with common interference problems, but can present a serious validation challenge when it does present itself. The 'masking' phenomenon involves proteins or peptide-containing drug compounds that bind endotoxin and thus reduce or prevent the ability of LAL to detect contaminating endotoxin and/or the added CSE spike which determines the sample's interference profile.

Pretreatment with an appropriate protease can cleave protein containing drug compounds and thereby make them unable to mask subsequent spikes. In some cases a compound's molecular shape can be modified with another protein (such as purified Bovine Serum Albumin, BSA) prior to exposure to CSE so that it no longer binds endotoxin, thus allowing for proper spike recovery. Indeed, one such compound tested in the Lilly lab appeared to be susceptible to proteolytic cleavage. However, after further study it was found that a more elegant solution was to use BSA to modify the binding capacity of the compound to allow for very good PPC recoveries. The defect associated with most proteases is twofold. First they are typically contaminated with endotoxin which is difficult to remove. Second, they are constituents of the LAL cascade and therefore likely to interfere with (falsely initiate) the LAL cascade if not removed prior to LAL addition. Fortunately most proteases are readily inactivated by heating in water at 80°C for 10 minutes. Perhaps, LAL manufacturers will provide purified proteases for such purposes in the future.

After many non-protease attempts, Petsch et al (1998) employed several different proteases to overcome the observed masking of protein compounds (ie to 'demask' the protein prior to testing with LAL). Treatment under a variety of conditions using Trypsin, Chymotrypsin, Pronase, and Proteinase K were all tried, with most success occurring with Proteinase K. Proteins used to examine the phenomenon included Lysozyme, Rnase A, Human IgG, bFGF, BSA, and Murine IgG1. Tests were done in pH 7, 20 mM phosphate buffer at protein concentrations of 1 mg/mL. The results of an initial and subsequent tests performed by passing a culture filtrate of *E. coli* (to mimic endogenous endotoxin) through a filter with and without each protein solution are shown in Table 3.3. The recoveries obtained demonstrate the amount of endotoxin that was bound by the protein and thus not recovered from the filtrate.

Table 3.3 (Petsch et al, 1998)

Protein	EU/mL without protein	EU/mL with protein
Lysozyme	6180	297
Rnase A	726	146
Human IgG	6180	99
b-FGF	478	9
BSA	6180	6100
Murine IgG1	6180	5840

The study demonstrates that the proteins with a net positive charge result in significant loss of endotoxin via binding, whereas those with a net negative charge do not. Significantly, the dilution of each of the above samples prior to filtration employed only protein concentrations that were non-interfering with the LAL assay, thus 'the reason for the poor endotoxin recoveries' are due to masking, and cannot be overcome by dilution.

KNOW WHEN A METHOD IS OPTIMISED (USING POLYNOMIAL REGRESSION)

The mathematical treatment of the kinetic standard curve has been accomplished via linear regression (LR) and also more recently by polynomial regression (PR). From the user's viewpoint the difference between the two is that the former draws a straight (best fit) line through the curve points, whereas the latter connects the points. Therefore, depending upon where on the line your sample result falls either treatment of the data could potentially be 'better' or 'truer' in providing the corresponding result. However, one can unequivocally surmise that for gauging the mid-point of the standard curve the polynomial result will be 'truer'. This can intuitively be seen in that the line touches the middle point using PR and often does not with LR (depending on the distance of r from 1.000). Thus if one wants to see if interference has been removed from a given test in method development, then look at the PR results.

This can be seen in a real life example. An analyst runs a standard curve in LAL reagent water (LRW) and runs the same LRW as a 'sample' with and without CSE (PPC). The analyst gets a good coefficient of correlation (r) and excellent %Cvs of <1.0%, however the %PPC recovery of the LRW 'sample' is 122%. Remember this is the same water and the same CSE used to make up the standard curve. Why, then, such a difference between the recovery of the mid-point of the standard curve CSE in LRW and the 'sample' LRW with CSE? A closer look at the data, however, reveals that the results are indeed the same. Both reacted in 1200 seconds. 1200 equals 1200 but 100% does not equal 122%. What is going on? Indeed, 100 may equal 122 by linear regression. This is because the outside curve points have distorted the inside curve, which does not touch the mid-point standard data point. Remember, in this example there can be no *true* inhibition or enhancement, because the 'sample' and the standard are exactly the same materials. This phenomenon is called the 'bow in the curve'.

If this was a development compound and not LRW, then one would be tempted to try and reduce the 122% PPC recovery to obtain more perfect results.

Any changes one could make to the test would only introduce interference to compensate for the 'bow in the curve' and thus lead away from the true removal of interference. A quick check of the PR result would show the user that nirvanah has indeed been reached, and the weapons of war against interference should be laid down to accept the test as it is.

There are, however, some caveats within the PR caveat itself. These include:

- the views of various international regulatory bodies in regard to the use of PR
- internal policy (one cannot use it arbitrarily to get the result one wants)
- the use of the PR analysis may change historical trends if incorporated.

In conclusion, the assay development question that began as 'what is the %PPC recovery?' ends up as 'Does the sample interfere with the absolute recovery of sample spike as determined by the sample spike recovery to the mid-point standard?' Or, more simply stated 'What test gives the best %PPC recovery using PR?'

APPRECIATING THE HISTORICAL DIRECTION OF THE USE OF LAL

Once in a while one hears someone nostalgically allude to the possibility of 'returning to gel clot'. Such a statement is often greeted by silent, vacant stares of those performing the testing. A hundred water samples a day by gel clot testing would be second only to a hundred rabbit pyrogen tests a day in terms of undesirability from a user standpoint. The advent of kinetic photometric testing, just as the advent of the gel clot test before it, brought about a many-fold increase in testing efficiency. It also allowed the spread of testing from the purely end-point test it had mostly been (with the exception of water) to the ability to test every phase of the pharmaceutical manufacturing process with greater accuracy. Such coverage of testing would simply not be practical by either the rabbit pyrogen assay or the gel clot assay. Similarly, the ability to extrapolate results against a standard curve brought about a refinement of the ability to quantitate endotoxin levels, making the invisible visible, and forming a comprehensive knowledge of contaminant control levels throughout the process. This task is still being expanded with the FDA's focus on Process Analytical Technology (PAT).

Modern pharmaceutical manufacturing processes include sampling and LAL testing of not only the finished (beginning, middle, and end of lot) and API material, but also in-process materials, including containers and closures, sterile water, bulk drug materials, and more recently, excipients and raw materials. The pyrogen assay included the housing of dedicated rabbits, and was therefore very expensive, and its expansion unlikely given cost and other resource constraints. The inability to quantify endotoxin associated with both pyrogen and gel clot testing acted as historical 'blind spots' restricting the improvement of processes now readily monitored, given the sensitivity and quantification associated with the LAL test. It is difficult to work toward lower specifications when performing an assay with an inherent invisible pass/fail result. Modern biopharmaceuticals may indeed contain trace amounts of endotoxin or may have activity (ie interferon) mimicking endotoxin. In such cases the accurate and reproducible quantification of these minute levels, as well as the differentiation of interference and endotoxin content, becomes paramount to demonstrating that allowable levels are present.

The first application of the clotting reaction discovered by Levin and Bang (Levin and Bang, 1956, 1964) was made by Cooper, Levin, and Wagner in their use of the 'pre-gel' to determine the endotoxin content in radiopharmaceuticals in 1970 (Cooper et al, 1970). According to Hochstein (Hochstein, 1982), Cooper was a graduate student at Johns Hopkins in 1970 and worked for the Bureau of Radiological Health. That summer Cooper persuaded the Bureau of Biologics (BoB) group led by Hochstein that a lysate from the horseshoe crab's blood would be useful in detecting endotoxin in biological products. Given the short half-life and stringent pyrogen requirements associated with radiopharmaceutical drugs, Cooper believed that LAL could be used to accomplish the improved detection of contaminated products. Though Cooper left the BoB to finish his graduate studies, Hochstein continued the Bureau's efforts to explore the use of LAL in the testing of drug products.

The potential for improvement in the area of pharmaceutical contamination control was evident in Cooper, Hochstein, and Seligman's very first application of the LAL test involving a biological (Cooper et al, 1972). The results of 26 influenza virus vaccines included as a subset of a 155 sample test using LAL varied from lot to lot by up to 1000-fold and revealed endotoxin in the 1 μg range (Cooper et al, 1972). Cooper later pointed out (Cooper and Pearson, 1976) that newer vaccines used in mass inoculation of Americans for A/Swine virus were subsequently required to contain not more than 6 ng/mL of endotoxin, a level that could not be demonstrated with pyrogen testing. Suspected adverse reactions were reported prior to the inception of the LAL assay and were an expected part of some drug reactions such as that associated with L-asparaginase antileukemic treatment as a product of *E. coli* (Oettgen, 1970). A third early application (radiopharmaceuticals and biological vaccines mentioned above) involved the detection of endotoxin in intrathecal injections (into the cerebrospinal fluid) of drugs. Cooper and Pearson reported that 10 such samples implicated in adverse patient responses were obtained, tested by LAL, and all 10 reacted strongly (Cooper and Pearson, 1976). The rabbit pyrogen test was negative for all samples when tested on a dose-per-weight basis. They concluded that the rabbit pyrogen test was not sensitive enough for such an application given that endotoxin was determined to be at least 1000 times more toxic when given intrathecally.

With regard to future direction the FDA has a vision of expanding the use of PAT to overcome the limitations of end-product testing. The trend in manufacturing is to know critical quality parameters throughout a process, and not rely on end-product testing, which is inherently probabilistic and provides a low level of assurance that a product has met the appropriate quality standards, including the lack of endotoxin. PAT includes on-line, in-line, and at-line measurement of manufacturing process parameters (Dziki and Novak, 2004). This philosophy change can be seen in comments made by the FDA Acting Deputy Commissioner for Operations:

> ... despite the slogan building quality in, most quality assessment today relies on end-product testing. This is a problem in and of itself ... we must turn to the science of manufacturing and the concept of quality by design (QbD), which means that product and process performance characteristics are scientifically designed to meet specific objectives, not merely empirically derived from performance of test batches.

LAL manufacturers are beginning to give users tools to perform PAT analysis, especially of water systems, in the PyroSense™ (Cambrex) and Endosafe® PTS (Charles River Endosafe) systems. The former is an automated in-line water test system utilising recombinant Factor C lysate and florescent reader. The latter is a rapid, hand-held portable device that allows testing without exhaustive user training on the spot (an at-line test).

AWARENESS OF THE POTENTIAL FOR NON-ENDOTOXIN PYROGENS

While endotoxin is of primary concern in traditional pharmaceutical manufacturing, given its stability, potency, and ubiquitous nature especially in water-thriving microorganisms, it is not the only microbial artifact capable of producing a pyrogenic response, especially given the rise of biotechnology and cell culture methods of manufacture. Therefore, two important themes relevant to the testing for pyrogens in drug products posit that:

1. Endotoxin is the most potent of the major microbial cellular residual artifacts and induces a wide range of deleterious host effects at the cellular and systemic levels. However, it is not the only one, or the only potent one. Cell wall constituents of gram-positive and gram-negative bacteria such as peptidoglycan and lipids are also capable of eliciting fever at greater doses.

2. Endotoxin activity can be increased by synergistic activity with non-endotoxin constituents (see Gentamicin episode below).

There have been several instances where pyrogens other than endotoxin have caused, either alone or synergistically with endotoxin, pyrogenic reactions in patients. A couple of instances are briefly discussed below.

Gentamicin

Perhaps the worst outbreak of pyrogenic parenteral product administered to patients in the US occurred between April 1998 and August 1999. Finished-product Gentamicin from two different US manufacturers which used the same API supplier contained a range of endotoxin activities up to the threshold pyrogenic response level (~5 EU/kg). Subsequent back-testing of some of the lots revealed that 'a significant number of patients developed pyrogenic reactions to doses as low as 2 EU/Kg' (Cooper et al, 2002). Thus it was suspected that, in addition to endotoxin, a non-endotoxin contaminant may also have been present acting synergistically.

The bulk supplier used the same process and process controls to supply both the dry and parenteral forms of the drug. Fanning and coworkers reviewed the Gentamicin adverse drug reactions (ADEs) (Fanning et al, 2000). There were 210 reactions involving 155 patients (multiple reactions per patient) who experienced chills, fever, respiratory symptoms, and shivering within three hours from the start of infusion. Reactions were nine times more likely associated with once daily dosage (ODD) therapy. Most of the events caused no serious complications, but five (3%) were severe, requiring intensive-care support. The series of events leading to the exposure of the contamination were summarized by Fanning (Fanning et al, 2000):

1 Multiple problems in production of the API including endotoxin contamination.

2 Use of ODD effectively unmasked the API problem.

3 Issue (2) led to unexpected adverse drug events due to higher concentration of impurities

In terms of 'learning points' associated with the event, the pyrogenic reactions that occurred revealed a few salient points:

- synergistic contaminants pose a risk of potentially lowering the threshold pyrogenic response, K, if they occur
- the episode suggests a wide range of responses associated with the human pyrogenic reaction (this can also be seen in human toll-like receptor (TLR) polymorphisms)
- the threshold pyrogenic response may truly be viewed as a limit and not a worst case level of pharmaceutical content, as some reactions can indeed occur at, or just below, the 5 EU/kg level
- accepting endotoxin levels which are clearly out-of-trend and near the endotoxin limit of the finished drug product should not be acceptable practice.

Baxter

Solutions of USP grade dialysis solution that had passed LAL testing for endotoxin showed pyrogenic activity when administered to patients between September 2001 and January 2003 (Martis et al, 2005(a); 2005(b)). A global recall was issued by Baxter Healthcare in May 2002. At this time Martis and colleagues set out to determine the cause(s) of the pyrogenic activity. The final result would be a dose-response curve established between the drug solution containing peptidoglycan and the peripheral blood mononuclear cell (PBMC) test which utilised IL-6 as the cytokine of choice for detection purposes.

In general terms a peritoneal dialysis system and associated materials ...:

> ...is a device that is used as an artificial kidney system for the treatment of patients with renal failure or toxemic conditions, and that consists of a peritoneal access device, an administration set for peritoneal dialysis, a source of dialysate, and, in some cases, a water purification mechanism. After the dialysate is instilled into the patient's peritoneal cavity, it is allowed to dwell there so that undesirable substances from the patient's blood pass through the lining membrane of the peritoneal cavity into this dialysate ... The source of dialysate may be sterile prepackaged dialysate (for semiautomatic peritoneal dialysate delivery systems or 'cycler systems') or dialysate prepared from dialysate concentrate and sterile purified water (for automatic peritoneal dialysate delivery systems or 'reverse osmosis' systems). Prepackaged dialysate intended for use with either of the peritoneal dialysate delivery systems is regulated by FDA as a drug (CFR, 2005).

Peritonitis results from treatment due to either bacterial contamination or, in the case of 'aseptic' or 'sterile' peritonitis, due to microbial artifacts in the dialysis solution. Peritonitis manifests itself in symptoms including gastrointestinal disturbance (nausea, vomiting, diarrhea) and fever, and is typically bacterial in nature, evidenced by positive cultures of the dialysis solution. However, instances of non-microbial or aseptic peritonitis also occurs. Outbreaks of aseptic peritonitis occurred in 1977 and 1988 and were determined to be due to endotoxin contamination (Karanicolas et al, 1977; Mangram et al, 1988). Icodextrin containing peritoneal dialysate contains a polymer of glucose derived from the hydrolysis of corn-starch.

In the Martis investigation the researchers followed 'a standard aetiological approach' including the following steps:

(1) verify and validate if common clinical materials, raw materials, or their handling was consistent with the cluster of complaints

(2) establish whether the product or its raw materials for manufacture meet internal and external regulatory specifications

(3) search for potential organisms or (bio)chemical contaminants that might be associated with or derived from the chemical components used in the product ... (ie extraordinary contaminants)

(4) if a difference is identified, implement corrective and preventative action plan immediately (Martis et al, 2005(a), (b)).

Dialysate effluents were taken from nine patients after dialysate instillation and dwell, six using standard glucose-containing dialysis solutions (negative control), and two using icodextrin-containing solution associated with peritonitis.

The resulting solutions were analyzed for icodextrin metabolites, total protein, peptidoglycan and pyrogenic cytokines: IL-6, IL-1b, and TNF-α. Significant results, relative to this discussion, include:

- elevation in the IL-6 dialysate effluent from an afflicted patient (>5000 ng/L) versus a non-afflicted patient (59 ng/L)
- no significant difference in IL-1b, and TNF-α responses
- a rough doubling of the protein content in the complaint sample 23.6 vs. 12.5 g/L
- endotoxin test results were within the allowed limit (<250 EU/L)

- no increase in rabbit response in traditional rabbit testing

- the IL-6 results were not affected by pre-treatment with endotoxin binding polymyxin B

- the activity was present in the raw material prior to product manufacture.

Subsequent analysis of 321 recalled batches of the drug revealed that 41% of the batches tested were positive for peptidoglycan (Cooper et al, 2002) as determined by the Silkworm larvae plasma test (SLP)*. The investigators found that of the two raw material suppliers, all affected batches were from the same supplier. PG concentrations ranged from the 7.4 μg/L detection limit up to 303 μg/L. Absolute determination of PG was precluded by the low concentration and presence of large molecular weight glucose polymers (matrix) that proved to be interfering. However, the presence of muramyl dipeptide, the smallest polymeric subunits of PG, were identified. The authors state that 'Because β-glucan is positive in both the silkworm larvae plasma and Limulus amoebocyte lysate tests, and the dialysate contaminant was positive only in the silkworm larvae plasma test, β-glucan was probably not the immunological provocateur'. However, it should be noted that LAL is a poor predictor of β–glucan content and β-glucan specific assays now exist.

Given the group's findings, they determined to identify the microbial culprits that caused the contamination. The production of icodextrin from cornstarch involves heat and acidification. Given such non-conventional culture conditions the group was able to isolate a thermophilic, acidophilic, gram positive organism: *Alicyclobacillus acidocaldarius* (previously *Bacillus acidocaldarius*). This is a common organism originally isolated from an acidic creek in Yellowstone National Park (Eckert and Schneider, 2003) and known to the food industry as a contaminant of orange and other juices and possessing a cell wall consisting of 40% peptidoglycan. The vegetative cells are easily destroyed by pasteurisation, however, they are spore formers and the spores persist (Alfredo et al, 2000).

The major learning points associated with this instance of contamination with non-endotoxin pyrogen include:

1. If a pyrogenic reaction occurs clinically and it is suspected that a nonendotoxin microbial pyrogen may be involved, then a monocytic test using IL-6 may be necessary

---

* Wako Pure Chemical Industries Co Ltdk, Osaka, Japan

2. The non-endotoxin microbial contaminant culprits may include a wide range of microorganism byproducts including peptidoglycan, exotoxins, protein A, lipoteichoic acid, glycoproteins, cell-surface polysaccharides, and DNA.

3. Novel standards and assays may be employed to preclude contamination in specific manufacturing processes in conjunction with regulatory oversight.

APPRECIATING THE RELEVANCE OF LIMULUS (AND OTHER ARTHROPODS) IN MEDICAL SCIENCE

This odd creature — a living fossil — seems so alien to us. However, with the recent advent of genetic comparisons of various divergent extant organisms it has become apparent that many features pertaining to the innate immune and developmental regulatory systems of *Limulus* and *Drosophila* are shared by virtually all multicellular creatures in various degrees (unfolding to researchers as a chronological gradient). This helps to understand *why* the blood of an arthropod proves such a useful tool in endotoxin detection.

Limulus lives in the soup of prokaryotes of the shoreline and *Drosophila* in the juices of rotting fruit. Both are front-line soldiers in the metazoan war against prokaryotes. Both use biosensors to detect endotoxin and an associated TLR to release their defensive protease-based defence cascade and antimicrobial defence molecules. Higher animals and man have developed extreme variations on this theme, with man possessing 11 known TLRs found to date (very close in number and sequence to the mouse) that detect everything from endotoxin and peptidoglycan and b-glucans to double stranded RNA and flagellin. Thus creatures are able to detect and destroy invading prokaryotes based upon detection of unique biochemical markers, including LPS, that do not exist in higher organisms (including *Limulus*).

Furthermore, in a surprising series of discoveries it has been found that virtually the entire lineage of multicellular life has been fostered from a small set of repeated regulatory genes (called HOX) that shape the anterior–posterior (head to tail) and dorsal–ventral (side to side) development of the embryo (see Figure 3.7). These genes have changed over time and in cases duplicated via duplicated chromosomes to add the complexity apparent today. However, all are built on a relatively simple genetic architecture (at least the commonalities and divergences between various representatives are simple to see, when pointed out by experts in the discipline).

Figure 3.7 Cartoon of arthropod and mammalian HOX developmental gene sequences. Boxes of DNA from the arthropod genome top have analogous sequences in the mammalian DNA boxes below dispersed on various chromosomes

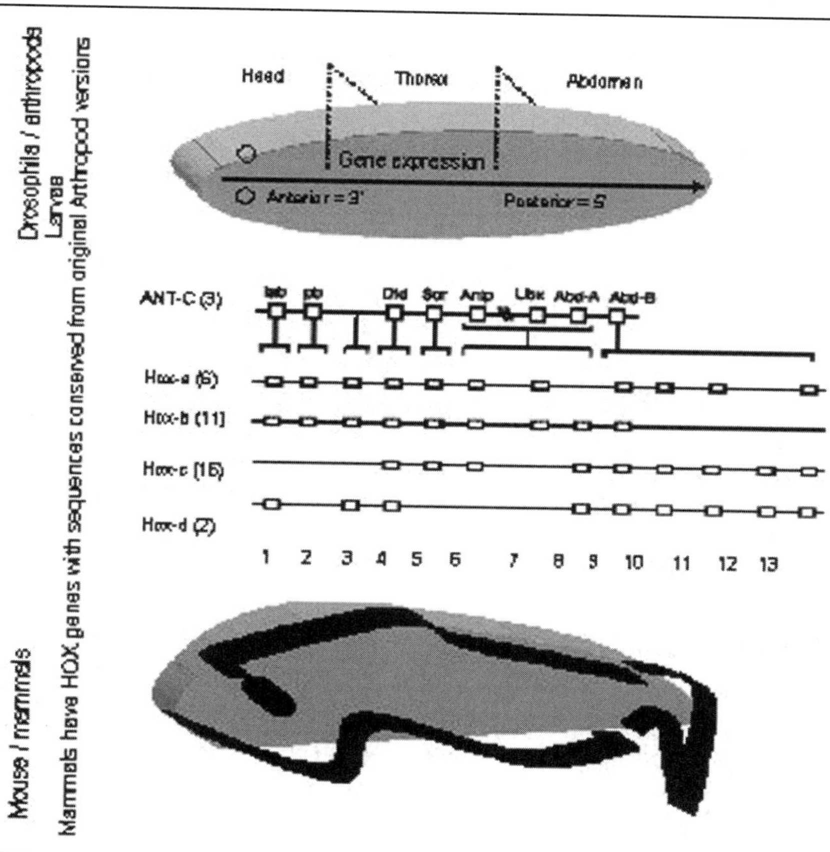

Bier and McGinnis (Bier and McGinnis, 2004) make much of the presence of a complex system of HOX genes in the ancestor of both arthropods and mammals. The criticality of this genetic conservation becomes apparent upon closer examination. Developmentally important genes have been phylogenetically conserved and disorders in humans (will) often involve genes controlling similar morphogenetic processes in vertebrates and invertebrates. A systematic analysis of human disease gene homologs in *Drosophila* supports this view — 75% of human disease genes are structurally related to genes present in *Drosophila* and more than a third of these human genes are closely related to their fruit fly counterparts. The authors present several cases of genes present in *Drosophila* representative of human disease-causing genes, including polyglutamine repeat neurodegenerative disorders. In this instance the size of the polyglutamine repeats can be related to the onset time and severity of the resultant neurological disorders in both flies and man. Indeed, the authors cite a very broad spectrum of shared mechanisms including CNS, cardiac, cancer, immune dysfunction and metabolic disorders that relate *Drosophila* genes to their human counterparts 'in virtually every known biochemical capacity ranging from transcription factors to signaling components to cytoskeletal elements to metabolic enzymes'.

In summary, Bier and McGinnis (2004) bring home the relevance of arthropoda, including *Limulus*, to modern biology:

> An important practical consequence of the fact that vertebrates and invertebrates derived from a shared, highly structured, bilateral ancestor is that many types of complex molecular machine which were present in this creature have remained virtually unchanged in both lineages ... These deep homologies between genetic networks can be exploited to understand the function of genes which can cause disease in humans when altered and should be very useful for identifying new genes in humans involved in disease states.

REFERENCES

Agalloco, J. (1992) Points to consider in the validation of equipment cleaning procedures. *J Parent Sci Tech* **46**(5), 163–168.

Alfredo, P. et al. (2000) Heat Resistance of *Alicyclobacillus acidocaldarius* in Water, Various Buffers, and Orange Juice, *Jour Food Protection* **63(10)**, 1377– 1380(4).

Avallone, H.L. (1986) Sterility Retesting *PDA Jour Sci & Tech* **40(2)**.

Bang, F.B. (1956) A bacterial disease of *Limulus Polyphemus*. *Bull Johns Hopkins Hosp* **98**, 325–351.

Bier, E., McGinnis, W. (2004) Model organisms in the study of development and disease, in Charles J. Epstein (ed) *Inborn Errors of Development*, Oxford Press, Vol. 49.Blumenthal, Robert, (0000) Cambrex, Walkersville, MD, personal communication.

Bruch, C.W. (1993) Quality Assurance for Medical Devices, in K.E. Avis, H.A. Lieberman, and L. Lachman (eds) *Pharmaceutical Dosage Forms: Parenteral Medications*, Dekker: New York. p. 487–526.

CFR (2004). Code of Federal Regulations, April 1, 2004, Subpart E — Control of Components and Drug Product Containers and Closures, S211.80–S211.115.

CFR (2005) Code of Federal Regulations, Title 21, vol. 8, revised April 1, 2005 (21 CFR 876.5630), Subchapter H — Medical Devices.

Cooper, J.F. (1990) Resolving LAL Test Interferences, *Jour Parenteral Sci & Tech* **44**(1), Jan–Feb, pp. 13–15.

Cooper, J.F. (2004) in Williams (ed) *Microbiological Contamination Control in Parenteral Manufacturing* Marcel Dekker.

Cooper, J.F., Brugger, P., Fanning, M., Matsuura, S., Poole, S. (2002) Alert and Action Endotoxin Levels for APIs: A collaborative study of pyrogenic gentamicin, *Parenteral Drug Assoc Annual Meeting* 11 December 2002, New Orleans, LA.

Cooper, J.F., Hochstein, D.H., Seligman, E.B. (1972) The limulus test for endotoxin (pyrogen) in radiopharmaceuticals and biologicals. *Bull Parenteral Drug Assoc* **26**, 153–162.

Cooper, J.F., Levin, J., Wagner, H.N. (1970) New rapid *in vitro* test for pyrogen in short-lived radiopharma-ceuticals. *J Nucl Med* **11**, 310.

Cooper, J.F., Pearson, S.M. (1976) Detection of endotoxin in biological products by the limulus test. In *International Symposium on Pyrogenicity, Innocuity and Toxicity Test Systems for Biological Products*, Budapest.

Dziki, W., Novak, K.J. (2004) Strategies for Successful Implementation of PAT in *Pharmaceutical Manufacturing, American Pharmaceutical Review* **7** (6), Nov–Dec, pp. 54–60.

Eckert, K., Schneider, E. (2003) A thermoacidophilic endoglucanase (CelB) from *Alicyclobacillus acidocaldarius* displays high sequence similarity to *arabinofuranosidases* belonging to family 51 of glycoside hydrolases. *Eur J Biochem* **270**, 3593–3602.

Fanning, M.M. et al. (2000) Pyrogenic reactions to gentamicin therapy, *New England Journal of Medicine* **343**(22), 1658–1659.

Hochstein, H.D. (1982) Review of the Bureau of Biologic's experience with limulus amebocyte lysate and endotoxin. In *Endotoxins and their Detection with the Limulus Amebocyte Lysate Test* Alan R. Liss: New York, 141–151.

Karanicolas, S. et al. (1977) Epidemic of aseptic peritonitis caused by endotoxin during chronic peritoneal dialysis. *New England Journal of Medicine* **296**: 1336–1337.

Levin, J., Bang, F.B. (1964(a)) The role of endotoxin in the extracellular coagulation of limulus blood. *Bull Johns Hopkins Hosp* **115**, 265–274.

Levin, J., Bang, F.B. (1964(b)) A description of cellular coagulation in the limulus. *Bull Johns Hopkins Hosp* 337–345.

Madsen, R.E., J., (1994) US v Barr Laboratories: A Technical Perspective. *PDA Jour Pharm Science & Technology* July–Aug. **48**(4): 176–179.

Mangram, A.J. et al. (1988) Outbreak of sterile peritonitis among continuous cycling peritoneal dialysis patients. *Kidney Int* **54**, 1367–1371.

Martis, L. et al. (2005(a)) Aseptic peritonitis due to peptidoglycan contamination of pharmacopoeia standard dialysis solution. *The Lancet* **365**(9459) 588–594.

Martis, L., et al. (2005(b)) Aseptic peritonitis due to peptidoglycan. *The Lancet*, **366** 289–290.

Munson, T.E. et al. (1993) Federal Regulation of Parenterals, in K.E. Avis, H.A. Liberman, L. Lachman (eds) *Phamaceutical Dosage Forms: Parenteral Medications* Dekker, New York, pp. 289–361.

Oettgen, H.F. et al. (1970) Toxicity of *E. coli L-asparaginase* in man. *Cancer* **25**, 253–278.

Petsch D., Deckwer, W.D., Anspach, F.B. (1998) Proteinase K digestion of proteins improves detection of bacterial endotoxins by the *Limulus* amebocyte lysate assay: application for endotoxin removal from cationic proteins. *Anal Biochem* 15; **259** (1): 42–47.

Rothschild, A. (1990) FDA Regulations and Guidelines, *PDA Jour Sci & Tech* **44**(1), Jan–Feb, pp. 26–29.

Sharp, J. (1995) Validation — How much is required? *PDA Jour Pharm Sci & Tech* **49**(3), 111–118.

US Dept of Health and Human Service, Food and Drug Administration (1987) 'Guideline on Validation of the Limulus Amebocyte Lysate Test as an End-Product Endotoxin Test for Human and Animal Parenteral Drugs, Biological Products and Medical Devices'.

USP (2000) Sterilization and Sterility Assurance of Compendial Articles, Chapter 1211.

Williams, K.L. (1998) Developing an Endotoxin Control Strategy for Parenteral Drug Substances and Excipients. *Pharmaceutical Technology Asia* Nov/Dec, special issue.

Williams, K.L. (1998) Developing an Endotoxin Control Strategy for Drug Products and Excipients. *Pharmaceutical Technology.*

Williams, K.L. (2001) *Endotoxins* Marcel Dekker, N.Y.

Williams, K.L. (2004). *Microbiological Contamination Control in Parenteral Manufacturing* Marcel Dekker.

4

THE ROLE OF THE QUALITY CONTROL MICROBIOLOGY LABORATORY IN THE CONTROL OF CONTAMINATION

Lucia Clontz

INTRODUCTION

The success of an Environmental Monitoring (EM) program is highly dependant on the performance of the Quality Control (QC) microbiology laboratory, including the proper training and skills of the microbiologists. This quality group plays a critical role in contamination control of the pharmaceutical manufacturing facility. The reliability of the data generated dictates the path taken by the company towards implementation of corrective and preventative measures to assure the quality and safety of products manufactured. Therefore the microbiology laboratory must maintain strict discipline in, and strict supervision of, its day-to-day operations. It is also critical for an open line of communication to be established between laboratory and manufacturing management so that timely communication of test results is achieved.

In many cases, the microbiology manager acts as an advisor to manufacturing management providing recommendations for facility design, facility and equipment cleaning/sanitization procedures, gowning procedures, proper clean room behavior as well as corrective actions in case of contamination events. For the laboratory to provide such crucial input into the company operations, laboratory personnel must be well versed in the relevant microbiological principles. Resulting from these many challenges and responsibilities, the QC microbiology laboratory sets itself apart from other QC operations and therefore, unique and especial managerial and technical training requirements are needed. These needs have been illustrated in the United States Pharmacopeia (USP) chapter <1117> *Microbiological Best Laboratory Practices*, published in the Supplement 2 of the USP29-NF24 (official date: August 1, 2006) and in the draft revision of USP chapter <1116>, *Microbiological Control and Monitoring Environments Used for the Manufacture of Healthcare Products* published in the PF Vol. 31(2) [Mar.–Apr., 2005]. These two chapters focus on personnel training. Other critical factors such as preparation and quality control of microbiological media, maintenance of microbial cultures, maintenance and calibration of equipment, laboratory records, and interpretation of assay results are also addressed. In addition to these two compendial documents, personnel training is addressed in the Parenteral Drug Association (PDA) Technical Report No. 35, *A Proposed Training Model for the Microbiological Function in the Pharmaceutical Industry* (published in November–December, 2001). Both the PDA and the USP have taken the lead to highlight the need for a broader and deeper understanding of microbiological concepts by individuals involved in drug product manufacturing, microbiological testing, and control of contamination in the manufacturing facilities.

In this chapter, the author addresses training for laboratory personnel, best laboratory practices, environmental monitoring and microbial identification programs, as well as technical and managerial skills requirements for QC microbiology management to ensure the success of a contamination control program in pharmaceutical and biotech manufacturing operations.

TRAINING OF QC PERSONNEL

Microbiological data are often subject to interpretation in relation to context, analysts' prior work experience, knowledge of microbiological concepts, as well as the technologies used to generate test results. Therefore, individuals involved in the performance and evaluation of microbiological testing must have the proper education, training, and experience to perform their jobs effectively. These individuals should be trained not only in the company's standard operating procedures (SOPs) which deal with good documentation and laboratory practices, but also have academic training in microbiology, aseptic technique, and laboratory investigations. Task training/on-the-job training is critical and therefore, EM operators and laboratory analysts must not be allowed to perform a test independently until they have demonstrated task competency. In addition, proper and detailed documentation of laboratory data is an area that must not be overlooked, given the lack of automation in most microbiological test disciplines and the need for human intervention in data collection and interpretation. It is also very important to emphasize to EM operators that they must be vigilant for potential contamination risks in production, and that they should document the activities in the manufacturing areas during sample collection. Such observations, to include number of operators in a given area and types of operations during sampling, can be crucial during investigations of EM excursions or anomalous test results.

Training for Clean Room Work

EM operators should receive additional training on the importance of good personal hygiene, aseptic techniques, disinfection/sanitization practices, plant gowning procedures, and proper clean room behavior. It is imperative that EM operators be made aware of their responsibility in reducing the risk of contaminating the manufacturing environment and the test samples while carrying out EM activities. They must understand that:

(1) natural human shedding is possible even while wearing appropriate clean room gowns

(2) sampling in critical zones must not be intrusive to avoid risk of product contamination

(3) both placement of equipment and personnel behavior can lead to adventitious product or sample contamination.

Besides microbial contamination, personnel also contribute to non-viable particle contamination typically associated with process equipment and materials. Non-viable particles can act as carriers for microbes shed by humans, and have the potential for contaminating work surfaces and product during manufacturing operations. Data on typical human shedding in a clean room environment are presented in Table 4.1 to illustrate the magnitude of personnel contamination. The data clearly show that even while wearing a two-piece coverall, operators can contaminate the work environment if they do not adhere to proper clean room behaviors. The data provided is further substantiated by studies which demonstrate that approximately 10^7 skin particles are disseminated into the air each day and that 10% of these contain viable microbes (bacteria, molds, and yeast) (Noble et al., 1965). In addition, it has been calculated that more than 500 species of bacteria live in the human mouth at any given time (Mayo Clinic, 2006) and these microbes can be introduced into the environment via talking, sneezing, or coughing.

Table 4.1. Non-Viable Particulate Personnel Contamination (Austin 2000)

Activity	Number of particulates (0.3μm) generated per minute by operators wearing a 2-piece coverall
No movement	4,000
Movement of hands, arms and neck	20,000
Sitting down or getting up	100,000
Walking at 2 mph	200,000
Walking at 3.5 mph	400,000

Therefore, it should not be surprising that excessive and unnecessary talking and rapid movements in critical zones can be viewed as inappropriate behaviors by the regulatory agencies. In fact, companies have received observations (FDA-483) and warning letters from the Food and Drug Administration (FDA) for improper behavior in aseptic areas (FDA, 2002, 2004, 2006).

Training in the principles of aseptic technique, potential sources of contamination, sanitization practices and clean room practices/personnel behavior should not be limited to EM operators and manufacturing personnel. In fact, this type of training should be extended to all personnel who enter a controlled environment, from Quality Assurance (QA) auditors and maintenance personnel, to visitors and senior management, so that the risk of contaminating the facility is minimized. Once a person enters a manufacturing processing area, he/she becomes part of the process with direct potential for product contamination. Strict adherence to gowning requirements must be observed, and once gowned, the integrity of gowns and gloves must be ensured. Hair and street clothes must not be visible, and if face masks are used, these cannot be removed during conversations. In summary, anyone who has access to controlled environments, must be properly trained to follow company SOPs and adhere to proper clean room behavior which includes appropriate speed of activities and movements.

Since people are the main source of contamination in a manufacturing environment, limiting the number of personnel during operations in a given suite is highly recommended and is often enforced by the regulatory agencies. Studies to establish maximum personnel load in manufacturing areas are typically performed during execution of air handling unit qualification protocols. EM is performed for air viable and non-viable particulates over a period of several days, under dynamic conditions, and with the proposed maximum number of personnel present in the room. If results obtained are below the set alert levels for the given room classification, the data generated support the proposed personnel load. Following such studies, the maximum number of people allowed in a given manufacturing suite during production is posted at the entrance of the rooms. Anyone with access to the production areas, including EM operators, must be trained and follow company procedures for personnel flow and room access to ensure they do not deviate from established company policies.

Training in Sample Collection, Preservation, and Storage

Proper sample collection must be emphasized during training and supervision of personnel performing EM and sampling for clean utilities, i.e., water, clean steam, gases, to ensure samples collected are not adversely affected. For example, collecting samples while room sanitization is in progress can lead to adventitious contamination or adulteration of the samples. In fact, it is not uncommon to detect low-level contamination of air-borne microorganisms from water samples due to inadequate sampling practices. The improper sanitization of water ports and valves or inadequate flushing of water ports can also lead to recovery of organisms that are not representative of the bioburden in the water systems. Sometimes, improper behavior when using a sanitizing agent can lead to false negative or false positive results. For example, a false negative bioburden result can arise from the splashing of a sanitizer onto an agar medium plate used for microbial recovery due to inhibition of microbial growth. False positive Total Organic Carbon (TOC) results are common due to the use of a sanitizer during sample collection by improperly trained operators — although TOC is a chemistry test, in many companies, the chemical and microbial evaluation of water and clean steam systems is the responsibility of the QC microbiology laboratory. Therefore, it is often the microbiologists skilled in aseptic technique who are responsible for sample collection for TOC analysis. If not properly trained, the operator may attempt to sanitize sample ports with sterile 70% isopropyl alcohol (IPA) prior to sample collection. This action, of course will most likely lead to false positive TOC results.

Training for Microbiology Analysts

Training for laboratory personnel who process EM and clean utility samples is also critical. Education in microbiology or related biological sciences should be a requirement for a person performing microbiological analysis in a QC laboratory. All laboratory analysts should receive technical training on aseptic techniques, disinfection/sanitization practices, sterilization procedures, growth promotion of microbiological media, in addition to the various microbiological techniques such as membrane filtration, pour- and spread-plate methods. Special technical training requirements must be in place for analysts responsible for microbial identification (ID) since even with basic techniques, such as gram staining, a higher level of expertise and careful attention to detail are needed to prevent misinterpretation of test results. In fact, many questionable ID results can be attributed to gram stains that are improperly performed (e.g., smears that are too thick or thin, over heat fixation, use of old cultures, and over decolorizing the specimens). If analyst technique is a concern, and additional training does not address the deficiencies observed, one option is to consider a semi-automated gram staining device to help eliminate some of the variability in the procedure. The Aerospray Gram Staining Instrument (Wescor, Inc. Salt Lake City, UT) and the manual Endosafe PTS™ Gram ID system (Charles River Laboratories, Wilmington, MA) have been successfully used by various laboratories and do help provide for more consistency in the gram staining techniques. However, even when automated devices are used in the testing laboratory, the technical expertise and experience of the microbiologists must be relied on for a correct evaluation of the test results.

Microbiology analysts must be trained to ensure strict adherence to aseptic techniques during handling and processing of samples. Although most microbiological testing is carried out in a high efficiency particulate air (HEPA) filtered environment, using a Biological Safety Cabinet (BSC) or Laminar Air Flow (LAF) hood, adventitious contamination during testing can still occur if analysts are not properly trained. To address such concerns, test negative controls and microbial monitoring of the testing environment (work surface, hood air, and personnel gloved fingers) are often performed. These quality control tests ensure suitability of the testing environment and personnel aseptic technique during testing. In addition, the data collected can further support the training status of the technicians and alert management when there is a need for remedial training in aseptic technique/behavior.

As with any other type of activity carried out in a regulated industry, analysts in the microbiology laboratory must receive training in, and show strict adherence to, good documentation practices which must be enforced by laboratory management. All analysts should also learn how to evaluate microbiological data, and how to perform laboratory investigations in case anomalous, out-of-specification (OOS), or out-of-trend (OOT) results are obtained. Personnel training records must be kept current for job functions, and management should require periodic performance assessments to ensure personnel have kept up with their required skill levels.

Training for QC Laboratory Management

In the pharmaceutical and biotech industries, there is an expectation that individuals who are responsible for the management of, and quality standards in, microbiological testing receive advanced training in microbiology, to include microbial physiology and taxonomy. Also, laboratory management must receive advanced training on principles of disinfection/sanitization, and sterilization procedures. Training in risk assessment/risk analysis is also recommended since this concept has been introduced to management of microbial contamination and microbiological data evaluation, and it is emphasized through the FDA's current Good Manufacturing Practice (cGMP) initiatives for the 21st century. Laboratory managers are directly responsible for investigation of sources of contamination and implementation of preventative and corrective measures and therefore must have adequate knowledge to make the right decisions and recommendations. For example, if an area becomes contaminated with spore-forming bacteria, it is imperative that a sporicidal agent is used to eradicate the contaminants since typical phenolic and quaternary ammonium compound products used for routine disinfection are ineffective against spore-forming organisms. However, recommending the use or overuse of a sporidical agent as routine practice for microbial control is clearly a decision based on lack of knowledge and will only lead to facility deterioration and equipment corrosion. Individuals in charge of contamination control must understand their power not only to control contamination, but also to cause damage to the facility, equipment, and products being manufactured.

In summary, education and training of the QC microbiology personnel as well as careful supervision of laboratory operations are key components of the overall quality system at a pharmaceutical company.

BEST PRACTICES IN A QC MICROBIOLOGY LABORATORY

The QC microbiology laboratory must have systems and programs in place to ensure the accuracy and reliability of the data generated, especially given the fact that there is an inherent variability associated with microbial testing which can impact the overall test results. For example, it is widely accepted that EM and water testing is at best semi-quantitative, since currently there is no one single test method that could recover all the viable microorganisms present in a given sample. Microbial recovery varies depending on the method (media and incubation conditions) and equipment used. It is not unusual to expect a 0.3 Log_{10} to 0.5

Log_{10} variability when performing microbial recovery techniques (USP <51> and <61>, 2006). In addition, many of the viable organisms present in the environment or in water systems are not culturable using traditional microbiological media and test conditions, and therefore cannot be detected. Additional testing variability can arise from the quality and sterility of media and materials used, sampling techniques, and the inappropriate choice of sampling sites and testing frequency. Thus, establishing sound quality control programs for the microbiology laboratory is critical to the success of the operations. Such programs include, but are not limited to, quality control testing of media and reagents used for EM and water testing, calibration and sanitization/sterilization of equipment, and identification of microbial isolates. The documentation system used by the laboratory, to include SOPs and test methods, is also critical and must be well designed and managed to ensure the quality of the data generated.

Laboratory Procedures

Microbiological test methods and SOPs that govern operations in the quality control microbiology laboratory must be clear, concise, and lack ambiguity, in order to prevent human errors and to standardize interpretation and reporting of test results. For example, the proper enumeration, documentation, and reporting of microbial colonies must be clearly defined in a procedure — some laboratories allow for a too-numerous-to-count (TNTC) result to be reported without further evaluation; others require estimation of count or testing of retained samples that are serial diluted so that a countable result is obtained. The possibility that a spreading colony or confluent growth could be incorrectly reported as TNTC may occur, thus directly impacting the interpretation and outcome of the test. Procedures must, therefore, provide clear guidance on how to report microbiological test results, including rounding of microbial counts to prevent analyst subjectivity and variability in data interpretation.

Quality Control Testing of Microbiological Media

Laboratory management often has to implement quality control testing programs that are cost effective, efficient, and yet compliant with the regulations — quality control testing of microbiological media is one such program. Whether the laboratory chooses to prepare the media in house or to purchase commercially-prepared media, testing for sterility and growth supporting properties must be performed to ensure the quality of the materials used for detection of microbial contamination.

According to the National Committee for Clinical Laboratory Standards (NCCLS) guideline, there are basically three regulations that apply directly to the quality control of microbiology media. The first one requires the user to demonstrate control over the shipment of materials, from the vendor to the user site — the user must document the *receipt and condition* of each batch of media received, and *notify the media manufacturer* of any of the following physical characteristics that could be indicative of exposure to undesirable conditions (excessive heat or cold, exposure to contaminants, or a breach of container integrity) during shipping: cracked Petri dishes, unequal filling of plates, cracked media in plates, hemolysis, freezing, excessive number of bubbles, and contamination.

The second and third QC requirements for media quality control pertain to *sterility and growth support* checks of the media. Typically, a representative number of media articles are incubated at specified conditions to confirm the sterility of the material. Growth promotion testing is performed to verify that a given medium is able to support growth of representative organisms and/or display the selective or inhibitory growth properties expected for the given challenge organism, whenever applicable — compendial organisms are often used in these challenge tests. For media used in environmental monitoring, growth promotion is performed using a wide spectrum of bacteria as well as fungal organisms (e.g., gram-positive cocci, gram-negative rods, gram-positive rods, spore-forming bacteria, yeasts and molds).

Although some compendial chapters call for quality control testing of every batch of in-house prepared or commercially prepared ready-to-use media, there are industry initiatives to reduce growth promotion testing to avoid redundant testing and reduce laboratory costs. According to the NCCLS, limited quality control testing of commercially prepared media is acceptable if:

(1) the laboratory maintains documentation (e.g., media label, package insert or brochure, invoice) that the quality control practices performed at the vendor site conform to specifications

(2) the laboratory maintains documentation of the receipt and condition of each batch or shipment of media, and of notification to the manufacturer of any problems with the media. In addition, media are exempt from growth promotion checks for every lot received if historical data, generated at the user site, indicate a failure rate of $\leq 0.5\%$, an accepted indication of proof of reliability. It is important to note that besides these specified requirements, the media manufacturer must be qualified and deemed an approved vendor by the user, as further assurance of the quality of materials received.

For in-house prepared media sterilized using validated cycles, growth promotion testing may be limited to each incoming lot of dehydrated media.

Indeed, the practice to reduce growth promotion testing by the user has been encouraged by many industry leaders in an effort to eliminate unnecessary, costly, and time-consuming duplication of quality control testing of microbiological media. However, the burden of proof is on the QC laboratory to develop the data, perform quality risk assessments, and request vendor records in order to justify their practices, and to prove the suitable quality of the media used in support of microbial testing. The use of temperature recording devices to monitor media shipments from vendor to user site is highly recommended by regulators and by the author in order to further ensure the quality of the materials received.

In addition to sterility and growth support verifications, special requirements are recommended for media used in EM testing — media plates/strips should be terminally sterilized and double bagged. If this is not possible, pre-incubation and 100% visual inspection of media articles prior to use is recommended. This practice is crucial for EM testing in critical zones to prevent the introduction of contaminants in the environment or a false positive result.

Maintenance of Stock Cultures

Stock cultures are critical lab standards and therefore the microbiology laboratory must have procedures that ensure they are carefully and consistently prepared, subcultured and stored. The laboratory must confirm purity, identity, and inoculum size (applicable for cell culture suspensions) upon receipt from the vendor, and as part of a routine quality control program. The use of a seed-lot technique, limiting the number of transfers to five passages* from the original culture, is widely used in the pharmaceutical industry and referenced in the USP (USP <51> and <61>, 2006).

Environmental isolates frequently recovered during routine EM and clean utilities testing should be maintained as part of the laboratory stock culture collection as reference organisms, and used to challenge the growth supporting capabilities of various types of media, including those used in environmental and water monitoring. Although this practice is encouraged by the regulatory agencies and has become standard in the pharmaceutical industry, one must realize that as soon as an organism is removed from the environment and cultured in the laboratory, it starts to lose its 'wild' phenotype. In fact, to adequately recover stressed organisms from environmental samples, microbiologists often have to subculture the isolates multiple times, thus significantly changing their behavior. Therefore it is imperative that maintenance of these types of cultures is well controlled so the environmental isolates used in media growth promotion testing

* One passage is defined as any form of subculture.

and in other challenge tests, such as method qualifications and disinfectant/sanitizer studies, will behave as closely as possible to their 'wild' counterparts. In many cases, the microbiologist must choose the best medium and incubation conditions thus ensuring the environmental isolate is maintained as close as possible to its 'wild' metabolic state. Although refrigeration is often used for preserving stock cultures, cryopreservation (–30°C or –70°C) is a preferred method for storage of environmental isolates since quite often an altered phenotype is observed after a single passage.

Handling of Microbiological Media, Materials, and Equipment

Careful handling of test media plates/strips is a must to prevent adventitious contamination leading to false positive results. Media plates used for environmental monitoring should only be removed from their protective bags in the environment where sampling will take place. Segregation of clean/sterilized and contaminated/non-sterilized materials is very important to prevent cross contamination. In most QC microbiology laboratories there is also a clear segregation of areas for live culture work (challenge tests and microbial identification) and aseptic work. In fact, the layout of testing laboratory areas has been subject to increased attention from inspectors during facility walk-thrus. When complete segregation is not possible, careful attention to sanitization and gowning practices must be observed. In addition, laboratory analysts must ensure test plates are protected (parafilm or clean plastic bags) when carried from one area to another. The concept of dividing clean/aseptic work areas (to include materials) from live culture areas also applies when preparing materials and equipment to be taken into manufacturing areas for environmental monitoring — EM operators should not take supplies and paperwork from an area of the laboratory dedicated to live culture work into manufacturing areas. It is also good practice to have dedicated EM equipment, carts, and supplies for the various classified areas in manufacturing. Equipment used for EM testing must be thoroughly disinfected and equipment parts that can be autoclaved should be sterilized. Materials used for microbiological EM and water testing must be sterile. The EM operator must observe personnel and equipment flow patterns and ensure that equipment and materials used in less clean environments are not taken into cleaner areas. Therefore careful planning must be in place when scheduling the EM and clean utilities sampling activities in the manufacturing areas.

Sample Handling and Tracking

EM operators and QC analysts must ensure sample integrity to prevent adventitious contamination or sample adulteration by following company procedures for sample collection, preservation, storage and chain of custody. For example, the types of containers used for water collection, as well as the manner in which samples are collected, stored, and tested become critical factors for an effective environmental and clean utilities monitoring program — samples can become contaminated during collection, transport to the QC laboratory, during storage prior to testing, or even during sample processing. Therefore it is imperative that sample collection and handling is performed by trained personnel and using procedures designed to prevent adventitious contamination.

After collection, samples should be processed promptly to avoid changes in microbial population. EM sample plates/strips must be protected while in transit to the QC laboratory, and incubated within a few hours of collection. It is not good practice to refrigerate EM sample plates/strips prior to incubation since this can delay growth of microorganisms during incubation. If not processed within one hour of collection, water samples must be stored under refrigerated conditions (2–8°C) and tested within eight hours of collection — for testing of coliform bacteria, a maximum holding time of 30 hours is allowed (APHA, 2005). The USP recommends processing samples within two hours of collection or storing under refrigerated conditions and processing samples within 12–48 hours of collection (USP <1231>, 2006). As indicated in USP chapter <1231>, *Water for Pharmaceutical Purposes*, samples not tested promptly after collection may be compromised — a decrease in bioburden can occur due to loss of cell viability or bacterial cell adhesion to the sample container. Conversely, an increase in bioburden can occur as a result of low levels of nutrients in the sample containers that could promote microbial growth. Since testing delays may occur from time to time, it is not uncommon for companies to perform studies to qualify a refrigerated sample holding time so that validity of microbial results from delayed tests can be assessed via the company deviation procedure and justified by supporting documentation. This practice is in fact recommended in USP chapter <1231> for determining "existence and acceptability of aberrations" in bioburden count as a result of delayed testing. One important point to consider is that storage conditions for microbiological samples must be designed to deter microbial proliferation while ensuring viability of organisms, especially of injured/stressed organisms which might be present in the sample.

Systems for sample tracking and submission into the QC laboratories must be well designed, documented and monitored. Sample containers must be labeled using indelible ink that will not be defaced when using a sanitizer, such as 70% IPA, or a disinfectant, such as a phenolic compound. Sample submission forms must include the type of sample collected, date/time of sampling, location from where sample was taken, sample quantity (if applicable), sample storage conditions, test method to be executed, and special instructions (if any) for sample handling. Upon arrival in the QC laboratory, analysts must document receipt and storage (location and conditions) of samples, record date/time of sample arrival, document the physical conditions of samples/containers and review sample submission forms for accuracy. The laboratory should also have procedures in place to control sample disposition after testing.

ESTABLISHING AND MANAGING MONITORING PROGRAMS FOR CLEAN ROOMS AND CLEAN UTILITIES

Along with a water system monitoring program, an EM program cannot be generic — both must be tailored to the facility and water system design. They must also be manageable and efficient so as to provide for test results that can be generated, reviewed and evaluated in a timely manner. These monitoring programs are tools to assist a company in the control of microbial contamination — adverse trends can indicate loss of microbial control. The choice of test methods, equipment, and sampling sites and testing frequency play an important role in the design of effective microbial monitoring programs, and therefore will be discussed in further detail.

Choice of Microbiological Media and Incubation Conditions

Selection of media and incubation conditions for environmental and clean utilities testing is critical so that a wide spectrum of organisms present in the environment and water systems can be detected. The bioburden of air, surfaces, and personnel in clean rooms is routinely monitored using a set of standard media that allow for detection of only a small fraction of viable microorganisms present in the environment. The traditional compendial method for bacterial recovery involves the use of an all-purpose medium such as tryptic soy agar (TSA) incubating at 30–35°C for a minimum of two to three days. Methods for recovery of fungi use a general mycological medium such as Sabouraud dextrose agar (SDA) incubating at 20–25°C for a minimum of five days. Media used for environmental surface sampling are typically supplemented with surfactants and neutralizers to overcome microbial inhibition that may arise from disinfectant residue on test sites.

Many companies chose to perform EM using TSA only with incubation at 30–35°C (Halls, 2004), since this medium supports growth of a variety of mesophilic bacteria and fungi that are common to pharmaceutical manufacturing environments. A dual-temperature incubation approach (either hot/cold or cold/hot) has also been used for EM purposes — a typical example of a hot/cold scheme is to incubate EM plates at 30–35°C for 2–3 days followed by incubation at 20–25°C for 3–5 days. Although the use of single medium/single temperature incubation approach (typically, 30–35°C for three days) provides benefits to an EM program in terms of costs and test turnaround times, regulatory agencies expect that site-specific studies be performed to prove that the chosen testing strategy, whatever it might be, is indeed the most appropriate for the environment being evaluated — in some cases, specific fungal medium and/or different incubation conditions may be required for adequate isolation of site-specific environmental isolates. For example, one study indicated that the use of TSA medium and a dual-incubation approach of 48 hours at 25° ± 2°C followed by 72 hours at 35° ± 2°C was needed to optimize the recovery of viable organisms, including fungi, from the manufacturer's environment (Reich, 1998).

In some cases it may be necessary to use alternate media and incubation conditions to recover certain types of organisms that might be of concern. For example, anaerobic media should be used for EM testing in cases where anaerobes (most likely found in spore form in the environment) or microaerophilic organisms are considered objectionable to a given manufacturing process or product, and/or when these types of organisms have been recovered from sterility tests. In addition, some companies choose to perform testing for detection of thermo-tolerant microbes in hot water systems, using appropriate media and incubation conditions.

Over the first few years of the 21st century many articles have been published and technical discussions held at conferences surrounding the need for detection of viable but non-culturable (VNC) organisms that can be present in pharmaceutical and biotechnology manufacturing environments and clean utilities. VNCs, many of which are oligotrophic microbes (also referred to as oligophilic microbes), live in extremely nutrient-poor environments and are often deemed unculturable by traditional methods. However, these types of organisms can be readily detected using methods that employ low-nutrient media and longer incubation periods, or when alternate technologies, such as ATP-bioluminescence, are employed. Some researchers have indicated that oligotroph counts exceed standard plate counts by up to two orders of magnitude. This can directly impact current microbial quality standards for controlled environments in pharmaceutical and biotech manufacturing areas which are based on traditional recovery of copiophilic organisms (also referred to as eutrophs or copiotrophs) that require a high-nutrient environment for survival (Nagarkar et al., 2001).

To generate supportive data for chosen test methodologies, companies often perform side-by-side studies using various types of media and incubation conditions. The test method that yields the best recovery (in terms of numbers and types of organisms) in the shortest period is then chosen for routine testing. This same approach applies when choosing the most suitable methodology for water testing — studies using TSA, R2A, and plate count agar are usually performed with various incubation conditions to choose the method that will detect most of the microbial flora in the water system in a timely fashion. Such studies are typically performed over a one-year period and can be time-consuming. However, the test results generated are valuable for justifying the methods chosen since data must be available prior to commitment to a given test method.

A company can spend a lot of time and money trying to recover all types of organisms present in a pharmaceutical manufacturing environment. However, from regulatory, business, and scientific perspectives, are all these additional/ alternate tests really needed? The answer is not straightforward and is often based on a risk management approach. QC management understands that although an EM program is critical to the overall microbial control program at a manufacturing facility, it has limitations that must be taken into consideration. In addition, as already mentioned, EM programs must be tailored to the facility design and type of production, and therefore will vary from company to company. So when choosing the methods for facility EM and water testing, QC microbiology management must consider testing costs, test turnaround times, and evaluate if the methods chosen will add value to the company's microbial control program. For example, studies have indicated that oligotrophic bacteria respond to typical disinfectants just as well as copiotrophic vegetative bacteria (Nagarkar et al., 2001). Therefore if these organisms are present in the environment, routine sanitization programs should be able to kill them or control their proliferation. Thus the decision not to test for oligotrophs/VNCs in the environment can be defended. However, if a particular type of organism isolated from a product sample was determined to require specific media and incubation conditions for detection, maybe, in this case, a company should consider supplementing EM and water monitoring programs with a method able to detect that particular organism, especially if the microbe has been deemed objectionable.

In summary, with the realization of the limitations to the current test methodologies for bioburden monitoring, companies should focus on looking for changes and trends in microbial contamination in a relatively short period. The timely detection of adverse trends and excursions is more critical to a microbial control program compared with an extensive testing regimen that attempts (unsuccessfully) to detect all organisms present in the environment.

Choice of Sample Sites and Testing Frequency

The manager of a QC laboratory is often involved in the selection of sample sites and testing frequency for the environmental and clean utilities monitoring programs at a pharmaceutical manufacturing facility. As with the choice of test method, site selection and testing frequency are critical factors for ensuring meaningful data are generated for the company's benefit.

During initial facility qualification studies, a large number of sample sites are chosen and frequently monitored against established acceptance criteria. Once the qualification studies are complete, reduced testing is carried out as part of a routine monitoring program. Based on the evaluation of the data obtained, the microbiology manager recommends test sites and selects representative microorganisms, recovered during qualification studies, to be maintained as reference organisms in the laboratory stock culture collection.

For routine environmental monitoring, it is critical to select sites based on room design, process activities and flow. Criteria used in site selection should include:

- sites prone to excursions during qualification studies
- sites close to critical zones/open process operations
- sites reflecting typical personnel and equipment flow.

The total number of sites selected for routine monitoring is not as critical as the rationale for site location. For example, monitoring a door knob in an area used for closed processes and where gloves are not required will most likely lead to detection of contamination by human-borne organisms (most often gram-positive cocci). These contamination events are often evaluated through time-consuming investigations, but no useful information will be gathered for a value-added corrective measure. In addition, the frequency of routine testing should be sufficient to allow for meaningful statistical calculations — generating too much data (sites and frequency) can lead to inefficiencies in the testing laboratory and untimely data review. On the other hand, infrequent testing and inappropriate choice of sites will generate data unsuitable for trend analysis or for a proper assessment of microbial control. Remember that the purpose of an EM program is to detect changes/trends in the environment and/or cleaning procedures that could pose a risk to product contamination.

Programs for testing water systems must also be carefully designed so that meaningful data are generated. As in the case of facility qualification, during initial water system qualification studies a large number of sample sites are chosen and frequently monitored against established acceptance criteria. Once the qualification studies are complete, reduced testing is performed as part of a routine monitoring program. Based on the evaluation of the data obtained, the microbiology manager recommends test sites and selects representative microorganisms recovered during qualification studies, to be maintained as reference organisms in the laboratory stock culture collection. For routine monitoring of water systems, it is important to incorporate a testing rotation of sample ports and use-points so the entire system can be monitored on a weekly basis. Also recommended is daily testing of supply and return ports and of use-points for water for injection (WFI) systems.

Choice of Equipment for Environmental Monitoring

An EM program includes active and passive sampling for air viable particulates, sampling for air non-viable particulates, as well as microbial monitoring of surfaces in the manufacturing areas, and personnel. Most surface sampling is performed using contact plates filled with TSA medium supplemented with Tween 80 and Lecithin or with Dey/Engley (D/E) neutralizing Agar (Difco). Many types of air samplers are available on the market — each having advantages and disadvantages. The QC microbiology laboratory manager must be knowledgeable about the different types of equipment and their applications when making a selection. When choosing the air sampling device, it must be ensured that:

- sampling activities will not contribute to contamination of the environment
- the equipment will be suitable for the monitored area
- the equipment can be calibrated and certified using National Institute of Standards and Technology (NIST) standards
- parts can be properly sanitized/sterilized to prevent adventitious contamination during sampling for microbiological attributes.

Once a piece of equipment is chosen for detection of air viable particulates, it is important not to change it without some type of evaluation, since microbial recoveries are different depending on the type of equipment used. Again, the purpose of an EM program is not to detect every microorganism in the environment but to monitor for adverse trends that can lead to risk of product contamination.

MICROBIAL IDENTIFICATION PROGRAM STRATEGY

The QC microbiology laboratory manager is responsible for developing a microbial identification (ID) program for isolates from manufacturing environments and clean utilities that will be both cost effective and compliant with the current regulatory expectations. Regulatory agencies do not expect companies to identify every type of organism detected — a costly and unnecessary approach. Organisms isolated from the environment and water systems are identified for trending purposes and to assist with manufacturing investigations in case of product contamination. Therefore a microbial ID program based on risk management and sound scientific principles should provide the data needed to ensure microbial control of the manufacturing facilities and clean utilities.

Alert and Action Level Excursions

When the number of detected organisms exceeds company established alert levels or limits, representative colonies are gram stained as part of an evaluation process into the possible source of contamination. Since the gram reaction must be observed using a microscope, this technique also provides for observation of cell morphology (i.e., rod, coccus, single cell, chain, cluster, etc.). Therefore the gram stain test can provide useful information as to possible source of contamination for a preliminary investigation. For example, most gram-positive cocci are human-borne and may be indicative of poor gowning techniques or improper clean room behavior. Most gram-negative rods are water-borne and may be indicative of wet conditions or standing water. Other simple tests such as spore staining, oxidase and catalase tests can be useful for a preliminary microbial identification. For most investigations into microbial excursions, these techniques provide sufficient information for data trending purposes.

When the number of recovered organisms exceeds company established action levels or limits, it is expected that an attempt is made to identify the isolated organism to the genus and species level for a better assessment of contamination source, risk to product contamination, and for implementation of effective preventative and/or corrective measures. Identification of typical microbial flora in the environment aids in the evaluation of the effectiveness of disinfection/sanitization and cleaning programs and personnel training, as well as early detection of deterioration of systems and facilities. Table 4.2 contains a summary of typical environmental contaminants in a pharmaceutical manufacturing environment and their associated potential sources and root causes.

Automated Microbial ID Systems

Many types of semi-automated and automated equipment are used for microbial identification. Microbial ID systems may be selected on cost criteria, database content (some systems have a greater representation of environmental isolates while others are designed with a greater representation of clinical isolates) and ease of use. To date, no single system is superior in terms of cost, data accuracy and precision, and database content. Some systems still do not have capability for mold identification; others have limitations in terms of certain groups of bacteria; and even the best systems still have difficulties with identification of stressed or viable, but unculturable organisms. Therefore managers of QC microbiology laboratories must take all these facts into consideration when deciding on a cost-effective and compliant microbial identification program for implementation in support of environmental and water monitoring activities at their facilities.

Some microbial ID systems are based on phenotypic profile of the unknown organism — others are based on genotyping. Phenotypic microbial identification is dependant on organism purity and age, metabolic state, as well as growth medium used to culture the isolate. Examples of automated phenotypical microbial identification systems include the VITEK Microbial Identification (bioMérieux Vitek, Durham, NC, USA) and the Biolog system (Biolog, Inc., Hayward, CA, USA). In the 1990s, systems for automated identification of microorganisms based on nucleic acid methodologies were introduced to the market. The MicroSeq TM 500 16S rRNA (Applied Biosystems, Foster City, CA, USA) and the Riboprinter (Dupont Qualicon, Wilmington, DE, USA) are examples of automated genotypic microbial identification and characterization systems that have proved both reliable and valuable when troubleshooting microbial contamination events.

As with equipment used for air monitoring, automated microbial identifications systems also have their strengths and weaknesses. Another important fact to consider is that different systems often generate a different genus and/or species for the same organism as a result of different technologies and changes in taxonomy. Reliance on one type of system as the primary tool for microbial identification, based on business needs and the value of the data generated, is highly recommended.

Phenotypic vs. Genotypic Microbial Identification

Phenotypic manual methods are less expensive but more time consuming. Automated phenotypic systems have the advantage of speed and automation of incubation and database analysis but are highly dependent on metabolic state of the unknown organism. For example, it is well known that for certain types of microorganisms, such as non-fermenting bacteria, reliance on biochemical profile is not the best approach for identification, since some types of bacteria exhibit little to no biochemical activity under given test conditions. Other organisms are too closely related to be adequately distinguished via phenotypic methods — examples include *Bacillus thuringiensis* and *Bacillus cereus* and many closely

Table 4.2 Pharmaceutical Environmental and Water System Contaminants

Type	Habitat	Likely Source of of Contamination	Safety Risks	Possible Root Cause of Contamination
Gram-positive cocci Genera commonly recovered: *Micrococcus*, *Staphylococcus* and *Streptococcus*	*Micrococci* are widespread in nature and are commonly found along with coagulase-negative *Staphylococcus* on the skin of humans. Gram-positive cocci typically represent a large percentage of the bacteria recovered from sampling indoor environments.	Personnel	*Micrococci* are not considered human pathogens *Staphylococcus aureus* and many species of *Streptococcus* are considered human pathogens	Inadequate aseptic and gowning practices Exposed hair and skin Inappropriate clean room behavior When isolated from water samples, often indication of poor sampling technique, insufficient flushing of sample port, or laboratory contamination.
Gram-negative cocci Most common genus recovered: *Neisseria*	Rare environmental contaminant Habitat is the mucous membranes of people	Personnel	Considered human pathogens	Inappropriate clean room behavior — no face mask, coughing, sneezing, excessive talking When isolated from water samples, often indication of poor sampling technique, or laboratory contamination.

Type	Habitat	Likely Source of Contamination	Safety Risks	Possible Root Cause of Contamination
Non-glucose fermenting gram-negative rods Genera commonly recovered include: *Alcaligenes*, *Ralstonia Acinetobacter*, *Burkholderia*, *Pseudomonas*, *Chryseomonas*, *Comamonas*, *Xanthomonas* and *Sphingomonas*	Natural habitats include plants, soil, intestinal tract of mammals, stagnant water sources as well as fresh flowing water. Typical environmental and water system contaminants	Contamination often associated with moisture in the environment and surfaces	Many are considered opportunistic human pathogens Bacterial endotoxins are a major concern if contamination is present in product path	Unattended stagnant water and wet floors Improperly drained drip pans, humidifiers, cooling towers and sink traps Floors that do not slope toward the drain When isolated from water samples, could be an indication of bioburden in the water system, valve, or sample port (inadequate sanitization or flushing). Adventitious contamination can occur in the laboratory during sample testing.

Type	Habitat	Likely Source of of Contamination	Safety Risks	Possible Root Cause of Contamination
Glucose-fermenting gram negative rods Organisms recovered from environment include: *Escherichia coli*, *Salmonella*, *Klebsiella* and *Enterobacter* species	Not typical environmental contaminants; sometimes recovered from water systems Wide variety of environmental niches Commonly colonize the human gastrointestinal tract	Contamination often associated with moisture, waste and sewage Recovered from fermentation and recovery areas where *E. coli* fermentation processes take place	Most are associated with human disease Bacterial endotoxins are a major concern if contamination is present in product path	Inadequate waste and sewage systems Inadequate cleaning/disinfection procedures post *E. coli* fermentation processes Inadequate personnel traffic pattern/feet contamination originating from restrooms When isolated from water samples, could be an indication of bioburden in water system, valve, or sample port contamination (inadequate sanitization).

Type	Habitat	Likely Source of Contamination	Safety Risks	Possible Root Cause of Contamination
Non-spore forming gram-positive rods Genus commonly recovered: *Corynebacterium*	Typical habitats include soil, food, plants and the skin and mucous membranes of mammals	Personnel Materials taken into controlled areas	Some are considered human pathogens	Inappropriate clean room behavior — no face mask, coughing, sneezing, excessive talking Inadequate sanitization of items taken into controlled areas When isolated from water samples, often indication of poor sampling technique, insufficient flushing of sample ports, or laboratory contamination
Spore-forming gram-positive rods Aerobic genera commonly recovered: *Bacillus* and *Paenibacillus*	Mostly saprophytic organisms Commonly recovered from soil, air, dust, debris and surfaces that come into contact with water	Materials taken into controlled areas Water fountains and condensation pans Floors, wheels, and feet	Majority of *Bacillus* species have little pathogenic potential; exceptions include *Bacillus cereus* (food poisoning) and *Bacillus anthracis*	Inadequate sanitization practices suggesting need for the use of a sporicidal agent Feet and wheel contamination

Type	Habitat	Likely Source of Contamination	Safety Risks	Possible Root Cause of Contamination
Spore-forming gram-positive rods (cont)				Damaged floors (cracks); floors that do not slope toward the drain. Unattended/unsanitized wet surfaces When isolated from water samples, often indication of poor sampling technique, insufficient flushing of sample ports, or laboratory contamination.
Actinomycetes Large group of filamentous and spore-forming bacteria (usually gram-positive) Main genera include: *Nocardia* and *Streptomyces*	Not common environmental isolates Widely distributed in soil and water Also isolated from barns, grain mills and air conditioning ducts	Materials taken into controlled areas	Most are strict saprophytes; some may become human parasites Associated with hyper-sensitivity pneumonitis and other allergic reactions	Poorly maintained air conditioning ducts/areas with no HEPA filtration Inadequate sanitization practices sugesting need for the use of a sporicidal agent When isolated from water samples, often indication of poor sampling technique, insufficient flushing of sample ports, or laboratory contamination.

Type	Habitat	Likely Source of of Contamination	Safety Risks	Possible Root Cause of Contamination
Yeast Most common genera recovered: *Candida*, *Cryptococcus*, and *Rhodotorula*	Not common environmental isolates Moist and dark environments Some are human-borne	Most commonly due to human-borne/personnel shedding Fermentation processes using yeast organisms (e.g., *Pichia pastoris*)	Some organisms are considered human pathogens	Inadequate aseptic gowning practices, exposed hair, skin, inappropriate clean room behavior Inadequate cleaning/ disinfection procedures post yeast fermentation processes When isolated from water samples, often indication of poor sampling technique, or laboratory contamination.
Filamentous fungi (mold) Most common genera recovered: *Aspergillus Penicillium*, *Paecilomyces*, and *Alternaria*	Moist and dark **environments,** soil, and plant material	Materials taken into controlled areas Moisture in the environment	Some organisms are considered human pathogens Concerns with production of mycotoxins if present in product path	Poor humidity control; presence of water damage and leaks. Areas with no HEPA filtration Inadequate sanitization practices suggesting need for the use of a sporicidal agent. When isolated from water samples, indication of poor sampling technique, or laboratory contamination.

related species of the genus *Staphylococcus*. Therefore application of nucleic acid-based methodologies eliminates some of the variables impacting biochemical testing, and provides for greater accuracy in test results. Currently (2006), many genotypic systems have limited databases since system libraries are still under development. As a result, unknowns processed using these types of systems may turn out to be a "no match." However this limitation can be overcome by genotypic profile of a "no match"/"unidentified" organism being given a code name and used to build up a system-customized database. In addition, the genotypic profile obtained can be used to perform a search for possible match using GenBank®, the National Institute of Health (NIH) annotated collection of all publicly available DNA sequences, accessible on the National Center for Biotechnology Information (NCBI) website.* GenBank® is part of the International Nucleotide Sequence Database Collaboration, which comprises the DNA DataBank of Japan (DDBJ), the European Molecular Biology Laboratory (EMBL), and GenBank® at NCBI. These three organizations exchange data on a daily basis and at the time of writing, there are approximately 59,000,000 bases in approximately 54,000,000 sequence records in the traditional GenBank® divisions and approximately 63,000,000,000 bases in approximately 12,000,000 sequence records in the Whole Genome Shotgun (WGS) division.

The fact that genotyping is a superior methodology for microbial identification is now widely accepted in the pharmaceutical industry. In the FDA guidance document on Sterile Drug Products Produced by Aseptic Processing (FDA, 2004), it is stated that genotypic methods have been shown to be more accurate and precise than traditional biochemical and phenotypic techniques. The USP Chapter <1117> also states that confirmation of identity of microorganisms should ideally be performed at the level of genotypic analysis. Genotypic methods are indeed more reliable and less subjective, but are also more technically challenging and expensive, and many in the industry believe their use should be limited to critical investigations associated with direct product failure. Fortunately, this is also the opinion of the some regulatory experts. For example, in the FDA guide for aseptic processing (FDA, 2004), it is stated that although genotypic methods can be valuable during product contamination investigations, appropriate biochemical and phenotypic methods can be used for the routine identification of isolates.

Although genotypic testing for routine microbial identifications is currently *not* a regulatory expectation, the author believes that this will no longer be the case in the near future. With advances in molecular biology and microbial ID

---

* *www.ncbi.nlm.nih.gov/Genbank*

technologies, phenotypic testing in general, especially as it relates to stressed environmental isolates and biofilm organisms, will most likely be viewed as an unacceptable practice. So, the microbiology manager must question whether the added accuracy and reliability in genus and species information can justify the costs associated with genotypic microbial identification systems.

Managing a Microbial ID Program

As already discussed in this chapter, microbial identification testing is costly and time consuming — therefore management must develop the most economical strategy that will comply with the current regulatory expectations and will add value to the environmental monitoring program. That does not mean that the laboratory needs to invest in different types of equipment for the various microbial identification needs. At a minimum, the laboratory should be capable of performing gram stains and simple biochemical tests typically used for preliminary testing to evaluate microbial contamination. Automated microbial ID systems are very useful and provide for time savings in the laboratory. However, since these systems can be quite costly, the QC microbiology manager must take into consideration the benefit of having an automated microbial ID system in comparison with manual methods, or subcontracting the work to an approved and qualified laboratory. Pharmaceutical QC microbiology laboratories may choose to invest in one of the automated microbial ID systems available on the market for greater control of processing time of routine isolates and for quality control testing of organisms in the stock culture collection. However, given the limitations of most systems, companies owning an automated microbial ID system still rely on contract testing laboratories for some of the identification work. It is also common practice for small pharmaceutical/biotech companies and non-sterile manufacturing sites to sub-contract all their microbial identification needs since they do not have in-house testing capabilities. One must consider that even when the ID work is contracted out, there is still a need for in-house microbiological expertise for evaluation and interpretation of the microbial ID results in the context of the microbiological control of the facility.

MICROBIOLOGICAL DATA MANAGEMENT

Evaluation and management of microbiological data are not easy tasks. Therefore they must be performed by qualified individuals who understand contamination control principles, the inherent variability and limitations of microbial testing, and the need for statistical analysis and trending of EM data — especially critical since microbial results are retrospective.

To ensure data reliability, management of the QC microbiology laboratory must routinely and promptly evaluate data generated to assess if there is an indication of inadequate training, improper behavior during sample collection, or a true adverse trend in the facility or clean utility system monitored.

Evaluation of Historical Data

Traditionally, companies have implemented alert and action levels/limits for EM and water systems data based on statistical analysis of historical data. Different methods have been used to perform statistical calculations, and some companies use software dedicated to analysis of environmental monitoring data. The use of a software program capable of trend analysis is extremely useful, and ensure excursions are promptly detected and the appropriate individuals informed in a timely manner — in fact, prompt and frequent review of EM data, with expeditious responses to adverse trends are regulatory expectations. The FDA has been known to cite companies for lack of diligence in the review of EM data (BioQuality, 2006).

Advances in microbiological control have led to small spreads between alert and action levels/limits resulting in little statistical significance. In addition, inherent variability in microbiological testing (PF, 2005) adds to the current dilemma when trying to react to EM excursions. In an attempt to address the lack of statistical differences in most trended EM and water data, the USP proposes evaluating EM data in terms of contamination incidence rate rather than numerical values established for alert and action levels (PF, 2005). The incident rate is defined as the rate at which environmental samples are found to contain any level of contamination — the USP recommends calculating mean incident rates for each clean room environment separately. Changes in the incident rate should be viewed as deviations from normal conditions. Thus, whenever the calculated incident rate trends above the recommended values for a given period, an investigation to assess root cause should be initiated. In taking this approach, companies should calculate their baseline incident rates and evaluate routine environmental monitoring data against those baseline values.

Microbial Deviations and Investigations

The microbiology QC manager is often involved with troubleshooting microbial excursions. This requires special training and knowledge in microbiology and, in many cases, statistics. The investigation plan created should be thorough so how and why the excursion occurred can be determined. It should also ensure an effective evaluation of any potential impact to product, equipment, facilities, and validated systems is performed. The investigation plan should include review of time of sampling, HEPA filter certification status, sanitization schedules, as well as room activities at time of sampling (e.g., room cleaning, maintenance work, and production activities). For microbial excursions, management should require identification of microbial isolates to assess potential sources of contamination. During the investigation, the EM operators should be interviewed for confirmation of appropriate sampling and aseptic techniques, and to asses whether any aberrant conditions were noted at time of monitoring. QC management should also review trended data for the area, and assess if prior similar excursions occurred which could give an indication that the corrective/preventive measures implemented were ineffective.

The laboratory testing environment and training of laboratory analysts should also be part of manufacturing EM and clean utilities investigations since there is always a possibility of laboratory error or improper sample handling. The probability of adventitious laboratory contamination increases as the criticality of the environment from where the sample was collected increases. Therefore, management should review media quality control testing results, preparation of standards, system suitability testing, test positive and negative controls, equipment calibration status, as well as sample handling and storage conditions.

Often the investigation plan prepared to troubleshoot microbial excursions will require additional sampling for a better assessment of the extent of the contamination, or to confirm an endpoint for an adverse trend. For example, for clean utilities, whenever suspect results are generated, evaluation of data from other sites in the system — especially upstream and downstream from the suspect port, as well as data from the supply and return ports (if applicable) — provide a good indication on whether the contamination is systemic, loop- or point-of-use related. As such, QC management must be familiar with piping and instrumentation diagrams (PI&D) for the clean utility systems in the facility. In case of EM excursions, the investigation plan may also require laboratory studies, using the organisms isolated from the environment, to challenge the disinfectant/ cleaning procedures. At the conclusion of the investigation, the QC manager will need to determine if the excursion was an isolated event, or part of an adverse trend requiring corrective actions. Some corrective measures may involve additional sanitization, use of alternate disinfectants, changes in gowning requirements, maintenance work, and/or verification of HEPA filter integrity.

Summary Reports for Trended Data

The QC laboratory is also responsible for preparing summary reports, containing data for EM and monitoring of clean utility systems, that will be reviewed by QA and manufacturing management. These reports are typically issued on a quarterly and/or annual basis and contain trended data presented in tabular and graphical formats, listing the types of organisms detected.

The monitoring programs of controlled environment and clean utilities in pharmaceutical and biotech manufacturing companies are often a focus during regulatory inspections — this is evidenced by the number of FDA citations given to companies for deficiencies observed in such programs (FDA, 2005). Therefore QC management must be aware of the latest regulatory expectations, and must ensure that trended data are reviewed in a timely manner, and that summary reports are prepared, approved by management, and available for review. Often the QC manager will be present during inspections to explain the data generated and to defend the company's EM program. Thus it is imperative that laboratory management:

- know about the latest industry trends
- are able to explain the rationale for setting alert/action levels/limits or incident rates
- are able to explain choice of sampling sites and frequency of testing
- can defend corrective actions implemented in case of EM excursions
- have the right skills to represent the department during inspections or audits.

TEAM WORK AND CUSTOMER SERVICE

The interaction between those in the QC laboratories and those individuals working in other departments within a pharmaceutical/biotech company directly impacts the success of the microbial control and environmental monitoring programs. The QC laboratory is an internal service organization supporting internal customers which include product development, manufacturing, cleaning validation and quality assurance groups. As such, QC management must understand and focus on internal customers, create a high-performance work environment, and ensure a culture of improved communication and innovation to meet customers' needs. The QC manager must also align resources to focus on what's important to the organization, and to solve microbial control problems using sound scientific methodologies. The tools used in Lean Manufacturing (Womack et al., 1996) could be used in a QC laboratory environment to ensure processes and activities deliver quicker customer value through improvement of sample flow and elimination of non-value added activities (waste). Lean is a comprehensive term referring to production methodologies and a management philosophy based on maximizing value and minimizing waste. In fact, any type of operation which requires interaction among various groups in an organization could benefit from the application of Lean tools. The Lean way of thinking fits very well in a QC microbiology laboratory environment, because it provides a way to do more with less human effort, less equipment, less time, and less space, while providing customers with exactly what they want. Those who work in microbiology laboratories can relate to this dilemma and appreciate this concept since microbiology work is very time-consuming, labor intensive and often performed with no to little automation. It seems that there is never enough staff or time to get the work done when the work needs to get done — so, if that is the case, the author's recommendation is to start thinking Lean!

Indeed, the job of a QC manager in the pharmaceutical and biotechnology industries can be extremely demanding but also very rewarding — managers are not only responsible for the success of laboratory operations, but often contribute to the design of microbial control programs in the facility. They must encourage team work among the various groups in the organization, and value the flexibility of the laboratory analysts and EM operators in accommodating the ever-changing manufacturing schedules. QC managers must be able to implement an effective planning strategy for all laboratory activities to ensure customers' needs are met in a timely manner. Streamlining laboratory process operations becomes critical to ensure a consistent flow of samples and results without bottlenecks and delays. Managers must line up all the operational steps and choose the best test methodologies that truly create value and deliver the results to manufacturing and quality assurance in the shortest time possible. The nature of microbiological testing and, quite often, the need for follow-up work for confirmation of test results, may prevent all steps in the sample processing flow from occurring in a desired timely manner. However, with good planning in place, provisions can be made for those extra tests so they do not create bottlenecks in the laboratory operations.

Good communication between the various groups in the organization and the QC laboratory is very important for the success of the day-to-day operations in a pharmaceutical/biotech company. There must be a shared understanding of the needs of the various departments which depend on the QC lab for test results, and of the needs of the QC laboratory itself. QC managers must plan for routine work and prepare for the unexpected (e.g., STAT samples, retests, etc.). A good approach to improving communication is to have an agreement about everyone's roles and responsibilities in writing — an interdepartmental service level agreement (SLA) is a useful tool to achieve just that. Figure 4.1 provides an example of an SLA that can be created between the QC microbiology laboratory and the manufacturing department.

Figure 4.1 Example of Service Level Agreement for QC Microbiology and Manufacturing

Agreement

QC Microbiology has committed to the timely reporting of test results and adequate response times for the EM of manufacturing areas. **QC microbiology** is also committed to notifying manufacturing whenever out-of-specification (OOS) or aberrant test results are observed and whenever laboratory investigations are initiated as they relate to production samples. **Manufacturing** has committed to providing sufficient notification for submission of production samples and request for non-routine EM samples. All production samples must be accompanied by sample submission forms that are filled out completely and correctly.

QC Microbiology Test Turnaround Time (includes testing and review)	Manufacturing Notification of Production Schedule Changes/Communication
Test / **Target Time** Bioburden: 3–5 days Endotoxin: 3 hours EM (routine): 5 days EM (special request): 5 days	• Provides 8-hour notice for sample submission and request for special EM coverage • Provides 24-hour notice for any change in production schedule that will impact sample submission or EM coverage to occur on 3rd shift or weekend • Ensures sample submission forms are accurate • Supplies a list of contacts as they relate to key production personnel on various shifts
Communication • Notifies nanufacturing within one working day in case OOS/ aberrant product test results • Supplies list of contacts as they relate to testing, EM shift coverage, and technical support	

Test	Target Time
Bioburden	3–5 days
Endotoxin	3 hours
EM (routine)	5 days
EM (special request)	5 days

Requirements/Constraints	Requirements/Constraints
• This agreement only applies to: scheduled production samples that are submitted to the laboratory from Monday through Friday between the hours of 0600 and 2000 • This agreement only applies to EM sampling that occurs during first and second shifts • There is no routine third shift or weekend shift to support product testing or EM	• Manufacturing must receive in-process test results within the target times • QC must provide EM support for all filling operations • On rare occasions, there may be delay (not more than 24 hours) in case a sample submission form needs to be corrected
SLA Contacts Print Name/Signature/Date	**SLA Contacts** Print Name/Signature/Date
Metrics to be Used to Monitor Agreement **QC Microbiology** will implement a tracking system to measure the attainment of target test turnaround times, EM response, sample schedule attainment by Manufacturing, and notification of schedule changes by manufacturing. **Manufacturing** will implement a tracking system to measure the attainment of target timely OOS/aberrant results communication by QC Microbiology, the number of times sample submission forms are returned for corrections and the number of occasions where delays in sample submission for corrections exceeded 24 hours. Metrics will be provided to the SLA contacts on a monthly basis.	

CONCLUSION

The role of the QC microbiology laboratory in the control of contamination in pharmaceutical and biotech environments is critical and one that must not be underestimated. QC microbiologists do more than just process samples and supply results to manufacturing and QA — they often participate in the bioburden sampling design for products, environment and clean utilities for the company. Whenever supplemental testing is needed, for confirming a test result, the microbiologist is often consulted for his/her expert opinion. Using sound science, he/she makes recommendations, weighing the cost of the additional work against the value of the additional data that will be generated.

Microbiologists are always caught in a battle between the need to report results in a timely manner and the inherent constraints in microbiological testing. Microbial contamination must be detected expeditiously, yet, traditional microbiological methods are time-consuming and often microbiological samples are perishable, leaving little room for testing of retains. Therefore, the QC manager must be kept up to date with current industry trends that could benefit the laboratory operations. One example is the array of rapid microbial methods currently being evaluated for compliance with the regulations and accuracy of testing — could any of these technologies be of value to the company?

The field of microbiology is changing fast — with the advances in molecular biology, it seems that old concepts on microbial physiology, recovery, taxonomy, and resistance to antimicrobials have been scrutinized, revised, and in some cases, completely changed. It is imperative that microbiologists receive special training in the most current technologies as well as on current regulatory and compendial testing requirements so they can continue to contribute to the overall success of the company's operations.

In summary, for a successful laboratory operation, management must ensure:

- everyone in the department understands their roles and responsibilities
- communicates frequently with their internal customers
- values the reduction of waste in the operations (non-value added steps/tests and rework).

A QC microbiology laboratory operation must be Lean to effectively meet customer needs while facing time, equipment, and staffing constraints in addition to the challenges of and unpredictability of microbial testing.

REFERENCES

American Public Health Association (2005) *Standard Methods for Examination of Water and Wastewater*, 21st Edition.

Austin, P., Ph.D., (2000) *Encyclopedia of Clean rooms, Bio-Clean Rooms and Aseptic Areas*.

BioQuality (2006) Vol.11(5) and (12).

FDA (2002, 2004, 2006) FDA.gov (reference: warning letters issued on 11/19/02, 03/21/04, 03/15/06).

FDA (2004) Guidance for Industry, Sterile Drug Products Produced by Aseptic Processing — Current Good Manufacturing Practice.

Halls, N. (2004) *Microbiological Contamination Control in Pharmaceutical Clean Rooms*, Boca Raton: CRC Press.

MayoClinic.com (2006) Oral health and overall health: Why a healthy mouth is good for your body.

Nagarkar, P.P. et al., (2001) Oligophilic Bacteria as Tools to Monitor Aseptic Pharmaceutical Production Units, *Applied and Environmental Microbiology*, **67**(3): 1371–1374.

Noble, W.C., Davies, R.R. (1965) Studies on the dispersal of *staphylococci. J. Clin. Pathol.*, **18**: 16–20.

PF (2005) USP chapter <1116>, Microbiological Control and Monitoring Environments Used for the Manufacture of Healthcare Products, Pharmacopeial Forum (PF) Vol. 31(2).

Reich, R. (1998) Environmental Contamination, Recovering Viable Environmental Particulates, *Medical Device & Diagnostic Industry Magazine.*

USP (2006) USP29-NF24 chapters <51>, Antimicrobial Effectiveness Testing and <61> Microbiological Examination of Non-sterile Products: Microbial Enumeration Tests.

USP (2006) USP29-NF24 chapter <1231>, Water for Pharmaceutical Purposes.

Womack, J.P., Jones, D.T. (1996) *Lean Thinking Banish Waste and Create Wealth in Your Corporation*, Simon & Schuster, NY.

5

RISK MANAGEMENT: PRACTICALITIES AND PROBLEMS IN PHARMACEUTICAL MANUFACTURE

Nigel Halls

'Risk' is the buzz word of the first decade of the 21st century in pharmaceutical manufacture and its regulation. This is because of Food and Drug Administration's (FDA) initiative *Pharmaceutical GMP's for the 21st Century — a Risk Based Approach* (FDA, 2004a). Older regulatory documentation places no emphasis on risk. For instance, the Code of Federal Regulations Part 211 *Current Good Manufacturing Practices for Finished Pharmaceuticals* contains no mention at all of 'risk', and although the EU good manufacturing practice (GMPs) mention the word several times, it is usually in the very general context of something like '3.17 in-process controls may be carried out within the production area provided they do not carry any risk for the production'.

The origins of FDA's initiative (FDA, 2004a) lie in its desire to make more efficient use of its resources without losing its focus on the protection of public health from actual and potential problems arising from manufacture of pharmaceutical products. FDA's own inspection programme of domestic pharmaceutical manufacturing sites is already being prioritised on risk analysis principles — attention is focussed on companies where inspections are considered likely to achieve the greatest public health impact, while at the same time the frequency and/or scope of inspections is moderated for other companies upon which FDA looks more favourably. Interestingly an analysis (Anon, 2006) has indicated that the number of GMP-related Warning Letters issued in 2005 dropped to half the number issued in 2004.

This is not only a US initiative. FDA makes it quite transparent (FDA, 2004a) that through increased participation with the *International Conference on Harmonization of Technical Requirements for Registration of Pharmaceuticals* (ICH), the *International Conference on Harmonization of Technical Requirements for Registration of Veterinary Medicinal Products* (VICH) and with the *Pharmaceutical Inspection Cooperation Scheme* (PIC/S) its intent is to bring risk analysis to all countries where pharmaceutical products are manufactured. ICH had by late 2005 (ICH, 2005) come close to completion of a Guideline — *Quality Risk Management, Q9* echoing many of FDA's sentiments.

In the words of Bob Dylan from over 40 years ago 'the times they are a-changin'.'

RISK AND WHAT IT MEANS

'Risk' is defined as the possibility of loss, injury or damage, alternatively as a dangerous element or factor (Anon, 1986).

It seems only reasonable that the pharmaceutical manufacturing industry should know where in its processes there may be *possibilities of loss, injury or damage*, or indeed *dangerous elements or factors*. This seems not only good sense from the perspective of protecting the patient, but also commercially.

It is also good sense, once a company knows where there are possibilities of loss, injury or damage, or dangerous elements or factors, that it should take some action to remove or minimise these risks.

One might think that adverse effects on public health arising from pharmaceutical manufacturing may have been more frequent and severe if the pharmaceutical industry were not already identifying and mitigating risk. This is however not FDA's thinking. FDA's perception appears to be that disaster has, up to now, only been avoided by rigid and unchanging compliance with prescriptive rules, and with limits agreed at initial product registration.

The balanced view is most likely that the pharmaceutical industry and its equipment suppliers have identified and mitigated manufacturing risk intuitively and probably erratically. FDA is pursuing a strategy which is likely to force a systematic, formal and more comprehensive approach. The FDA sees this as a means of accelerating the evolution of pharmaceutical manufacturing from an art to a science- and engineering-based activity (FDA, 2004a) — which must be a good thing.

Risk

Nothing is without risk. A vast global insurance industry thrives on risk. In this chapter risk is defined as *the possibility of something going wrong*.

Given this definition there are two important dimensions associated with risk which must be understood — *frequency of occurrence* and *severity*.

For example, the severity of risk from earthquakes in California is high, but the occurrence is low. Similarly, the frequency of occurrence of civil aviation disasters is low, but the severity is high. The frequency of occurrence of train services being late in the UK is relatively high, generally the severity is low.

The scales applying to what is meant by *high* and what is meant by *low* are specific to each case; in the context of the San Francisco earthquake of 1906, 2–3000 deaths is severe; in the context of civil aviation 2–300 deaths is severe. With respect to frequency of occurrence, a low frequency of civil air crashes would be one crash per 10 years, in the context of earthquakes in California a low frequency might be one in 200 years.

In the context of contamination control in pharmaceutical manufacture, the severity of viable microbiological contamination in aseptically filled parenteral injections would be high. On the other hand, the severity of contamination of Aspirin tablets by viable microorganisms would be lower than of contamination of Aspirin tablets by Warfarin. The concepts of severity and frequency of occurrence are illustrated in Figure 5.1.

Risk Analysis

In this chapter, risk analysis is used in the sense of identification of risk. Some of the techniques described as 'risk analysis tools' provide mcuh more than merely a list of risks.

Risk Management

Risk management is an 'umbrella' concept which takes in the understanding of risk, the identification of risk, and its subsequent mitigation.

Risk management should be understood to be both proactive and continuing. It is not a reactive problem-solving tool. Like validation, risk management should be undertaken before embarking on a new product, piece of equipment or process, and then repeated at intervals over its life cycle as new information, newer technologies and changed economic conditions emerge.

Figure 5.1 The Concepts of Severity and Frequency of Occurrence as Part of Risk Management

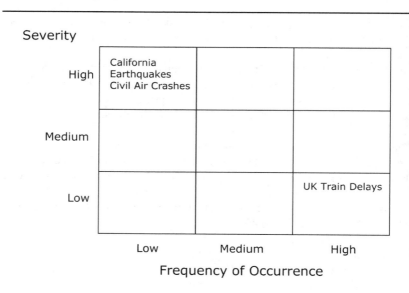

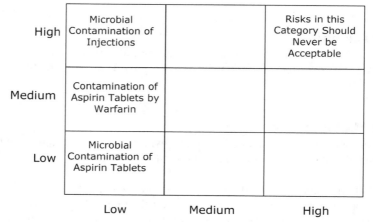

Severity	Low	Medium	High
High	California Earthquakes Civil Air Crashes		
Medium			
Low			UK Train Delays

Frequency of Occurrence

Severity	Low	Medium	High
High	Microbial Contamination of Injections		Risks in this Category Should Never be Acceptable
Medium	Contamination of Aspirin Tablets by Warfarin		
Low	Microbial Contamination of Aspirin Tablets		

Frequency of Occurrence

Although risk analysis is something most likely championed and operated by a company's technical management, the broader concept of risk management must be supported by those levels of higher management having financial control.

Risk Mitigation

Risk mitigation usually involves spending money in the short term to save money in the long term, and in the pharmaceutical industry perhaps to save lives too.

All risks are not equally consequential — risk mitigation involves making responsible justifiable decisions. Risks identified as being of low severity and with a low frequency are very likely to be tolerated on the basis of 'what cannot be cured, must be endured'. On the other hand, formal risk management techniques demand that any risks which have both high frequencies of occurrence and high severity are given immediate priority for mitigation (Figure 5.1).

Many risks will fall into neither of these extreme categories. Thus mitigation is about making an informed choice from the following options.

- Avoidance or elimination of risk by redesigning the product, process or piece of equipment. This is often the most costly option, not necessarily in monetary terms but certainly in time. A redesign may introduce new risks.

- Mitigation of risk by use of improved or alternative technology. For instance, isolation technology is now common-place in performing the *Test for Sterility* in pharmaceutical control laboratories. Without doubt the reason for this lies with the change introduced by the pharmacopoeias in the mid-1990s which eliminated automatic re-testing in response to 'first test failure'. Prior to these changed requirements the capital costs of introducing isolation technology could not be balanced against rejection costs for failing batches.

- Mitigation of risk by introduction of technological compromise. An example of a compromise is what is in the US called 'redundancy': EU GMP Annex 1.83 (Anon, 2002) reads 'due to the potential additional risks of the filtration method as compared with other sterilisation processes, a second filtration via a further sterilised microorganism retaining filter, immediately prior to filling, may be advisable'. Another compromise is use of 'fail-safe' principles, eg temperature overrides which shut down processes when temperatures get too high, and when the monitoring devices fail.

- Improvements in the awareness of risk or of drift towards risk by increased monitoring. The best way of doing this is obviously by continuous real-time monitoring linked to automated alarm systems. Manual monitoring begins to introduce the additional risks associated with human error.

- Mitigation of risk by product testing. In general industry practice, even where 100% non-destructive techniques are available, this is considered the weakest of all strategies for mitigation of risk. It is paradoxical that in pharmaceutical manufacture 100% non-destructive finished product testing is frequently viewed as a highly desirable ideal.

RISK ANALYSIS TOOLS

Risk management has been practiced for many years in industries other than pharmaceutical manufacture — various techniques (risk analysis tools) are available. Each of them was initially developed for a particular application, and consequently the strength of each technique resides in the application for which it was originally intended.

Application of risk management to pharmaceutical manufacture should sensibly start from knowledge of the risk analysis tools used successfully in other industries. It must be cautioned, however, that no risk analysis tool is without its own requirements for making reasoned decisions, and although it may well be possible to use an existing tool plucked directly from, say, the food industry, it may be more sensible to modify that tool to the particular requirements of pharmaceutical manufacture.

In this section, two risk analysis tools will be discussed in detail — *Hazard Analysis Critical Control Points* (HACCP) and *Failure Modes and Effects Analysis* (FMEA). Others are available but these two are the most strongly advocated, best known and most widely used. Both techniques share two important prerequisites which are themselves of such importance that they are given separate sub-headings within this section. These are assembly of the risk analysis team, and preparation of a process flow map or diagram.

The Risk Analysis Team

There is a saying 'it is the bad workman who blames his tools'. The output of the best risk analysis tool can be no better than the team of people using it. Risk analysis encompasses so many different skills that it can never be a solitary activity — it demands a team approach. Within the team are three important factors which should be considered as a minimum — the composition of the team, the leader (chair, facilitator, etc) of the team, and the technical secretary (scribe, note-taker, etc) of the team.

- *Team composition.* The 16/17th century English poet John Donne wrote that "No man is an Iland, intire of it selfe; every man is a peece of the Continent, a part of the maine'. So it is with any product, item of equipment or process — no single individual has complete knowledge of it. Successful output from risk analysis requires assembling a team which collectively has, as far as possible, a complete knowledge of the subject under analysis.

  This means taking membership from all disciplines who may be involved with the subject under analysis, and from different levels of the company's hierarchy. Practical knowledge of the day-to-day operation or maintenance of a process lies with the people who undertake this work. Knowledge of the consequences of, say, product variability to the patient may lie with medically or pharmaceutically qualified personnel. The microbiologist may have a thorough understanding of the principles of industrial hygiene. Important information may be held only by the formulation scientists involved in the product's initial development. What is important is that each team member should be *capable*, in the sense of actually having some knowledge to contribute to the analysis.

  It is also important to understand that participation in a risk analysis team places demands on people's time and energy. Personnel chosen should not merely be capable of contributing but should be *motivated* to contribute — in other words they should want their work to lead to a high quality output. Many factors contribute to motivation — curiosity is a motivator in any new venture, variety is another (many day-to-day roles are quite boring and repetitive). In a well-managed company, most people look favourably on, and are quite well motivated towards innovation — that impetus will be lost if the outputs from risk analysis teams do not lead to evident processes of risk mitigation.

  The composition of the team should therefore be capable of addressing the whole process, from its day-to-day operation to its broadest implications, and be drawn from personnel who are capable of contributing knowledge and information regardless of their status in the hierarchy of the company. Each individual should be motivated to contribute to a high quality output. At the same time, the numbers of participants in the team should be manageable — involvement usually optimises with around five to seven members and has noticeably started to 'tail off' as numbers reach double figures.

- *Team leader*. The quality of the output from any team, particularly from a team drawn from individuals with different backgrounds, is strongly influenced by its leadership.

  Team leadership and management of meetings can be taught. Risk analysis teams may be composed of both managers, who are very experienced at putting their points across in meetings, and maintenance engineers, who rarely, if ever, attend meetings. They may contain both production personnel with strong meeting skills developed through wage negotiation committees, and scientists, whose skills focus mainly on data analysis. Successful output depends on the team leader harnessing these disparate skill sets in the best interest of the subject undergoing analysis.

  Although these expectations of the team leader point towards a 'management consultant' approach to leading risk analysis, this must be ameliorated by appointing to each risk analysis a team leader who has some affiliation with the particular subject. For best use of risk analysis tools, the team leader must make technical decisions, both prior to and within the meeting. These require some knowledge of the product, item of equipment or process being analysed.

  The team leader need not be a technical expert on (say) autoclaves or High Velocity Air Conditioning systems, to lead a risk analysis on one of these areas, but he must be familiar with the subject and its vocabulary. He must recognise when meaning becomes drowned in words and have the ability to articulate and summarise meaningfully. Each risk analysis project may be budgeted for a few hours or a few days, according to its perceived complexity — risk mitigation will continue for longer. The team leader's role is to ensure that a practical end-point is reached in the time available.

  All team leaders should be trained in the use of the risk analysis tool they are using, and may benefit from the presence of their trainer as an observer or facilitator when first puts this knowledge into practice.

  In summary, risk analysis team leaders should be appointed individually to each subject presented for analysis, on the basis of having some knowledge or affiliation for the subject. In addition the team leader should have training in the risk analysis tool to be used, and have been trained, and have some experience in managing meetings and leading teams.

- *Technical secretary*. It is sensible to document the risk analysis — this will probably become a regulatory obligation. This is something which is relatively new to the pharmaceutical manufacturing industry so it should be understood that the way in which documentation is approached *de novo* may influence the output not only of individual projects, but of the overall value of risk analysis in the future. Documentation is important, but essentially secondary, to the technical effort it supports. The 'tail should not be allowed to wag the dog'.

As for the team leader, the technical secretary for a particular risk analysis subject should be conversant with the technical vocabulary of the subject. He should also be very familiar with the application of the risk analysis tool, as a critical part of his role is to ensure the analysis fits into the structure predetermined for the particular tool. This is more than just note-taking. If the team leader is 'piloting' the project, the technical secretary is 'navigating'.

Some risk analysis tools (for instance FMEA) have a set format for their documentation. These may be in hard copy, for ongoing note taking and re-typing afterwards. Computer software is available — this may speed up the overall process time from initial meeting to final output, approved and signed off, but care should be taken to ensure that its use in risk analysis meetings does not slow down the technical input.

In all risk analysis projects the team leader and the technical secretary should make some preparations prior to gathering the team and applying the risk analysis tool. For instance, it is sensible to have standard operating procedures, validation dossiers, materials specifications, engineering drawings, operating manuals, etc to hand before the team assembles. It may also be wise to have decided the boundaries of the analysis, and to have an outline process map available (see below).

Process Mapping

A process map (flow chart) is the second prerequisite of risk analysis. The text book process (see Figure 5.2) is a linear model in which starting materials undergo some form of conversion into finished products. Reality is more complex.

In any risk analysis a clear visualisation of the process and a clear delineation of the part of the process which is the subject of the analysis is necessary.

In the scientific sense, mapping the process is observational, in that it should represent what happens in the exact sequence that it happens. Many companies include flow charts and process maps in their SOPs — these may be helpful in

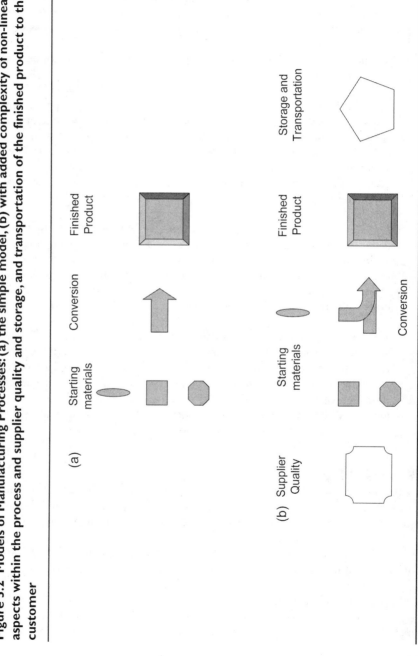

Figure 5.2 Models of Manufacturing Processes: (a) the simple model, (b) with added complexity of non-linear aspects within the process and supplier quality and storage, and transportation of the finished product to the customer

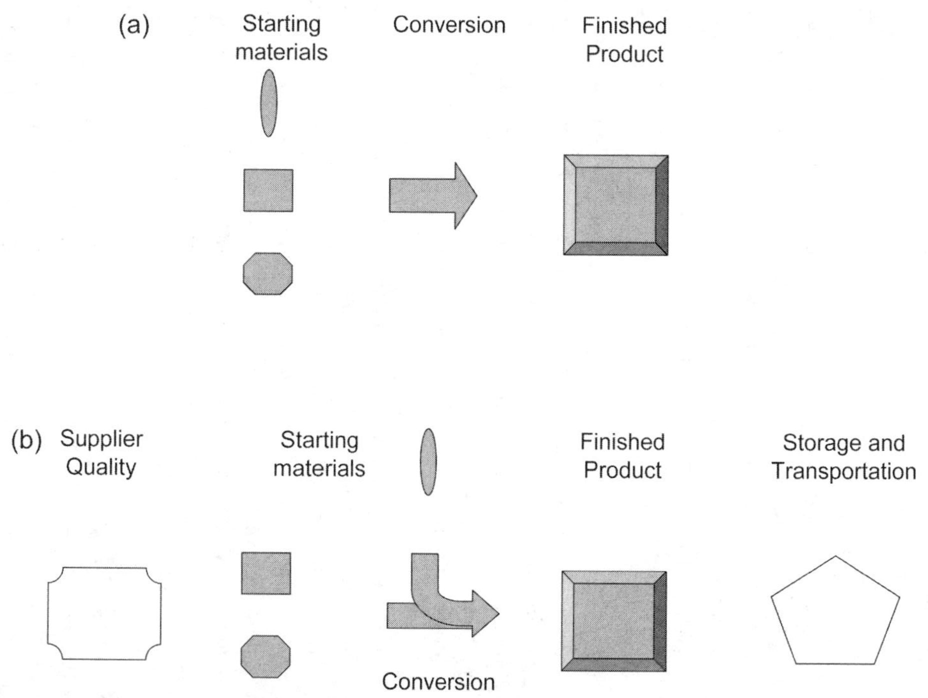

clarifying wordy descriptions and in identifying minor steps, subsidiary processes, essential links, etc, known to the direct workforce but invisible to office-bound management.

Figure 5.3 illustrates part of a process map for dispensing. Importantly it indicates two physical locations which must exist (either separately or as a single location) and should therefore be defined —

1. Somewhere where the warehouseman leaves bulk materials for the dispenser to collect (should the dispenser go to the warehouse, or should the warehouseman deliver to the dispensary?).

2. Somewhere where the dispenser stores bulk materials before taking them into the dispensary prior to weighing, and where he leaves excess materials after weighing out the requisite amounts.

Process mapping needs to be done at a level useful both to the direct workforce and to management — the process of dispensing could have been mapped in three stages (dispenser takes material, dispenser weighs material, dispenser returns excess) but with less value.

Most risk analysis tools work best with linear processes or with linear segments of more complex processes — these must be clearly delineated from the outset, or the risk analysis may possibly run out of control. Figure 5.2(b) is an expansion of the text book model of Figure 5.2(a) — one starting material is introduced only after some 'conversion' has been undertaken with the other starting materials, and the impact of supplier quality and storage and transportation of the finished product to the customer has been included. In using any risk analysis tool the team leader is at least responsible, usually in conjunction with the technical secretary, for drawing up a cursory process map before the risk analysis meeting. (The boundaries of the process to be analysed should have also been decided.) The team leader should be prepared to justify his decisions with respect to the process boundaries, and also be prepared to see the process map modified in light of the knowledge imparted from the risk analysis team.

Other clarifications may be achieved from process mapping in some applications. A process map showing (partially) how materials are moved from one graded area within an aseptic manufacturing suite to another, is illustrated in Figure 5.4. This clearly indicates the grade of area for each operation, and how materials are protected in transit from one grade to another. It also denotes where there are other SOPs, describing activities such as filter integrity testing, routinely carried out in such an operation, but not part of the linear process itself.

Figure 5.3 Partial Process Map of Dispensing Illustrating Essential Process Steps Prior to Weighing

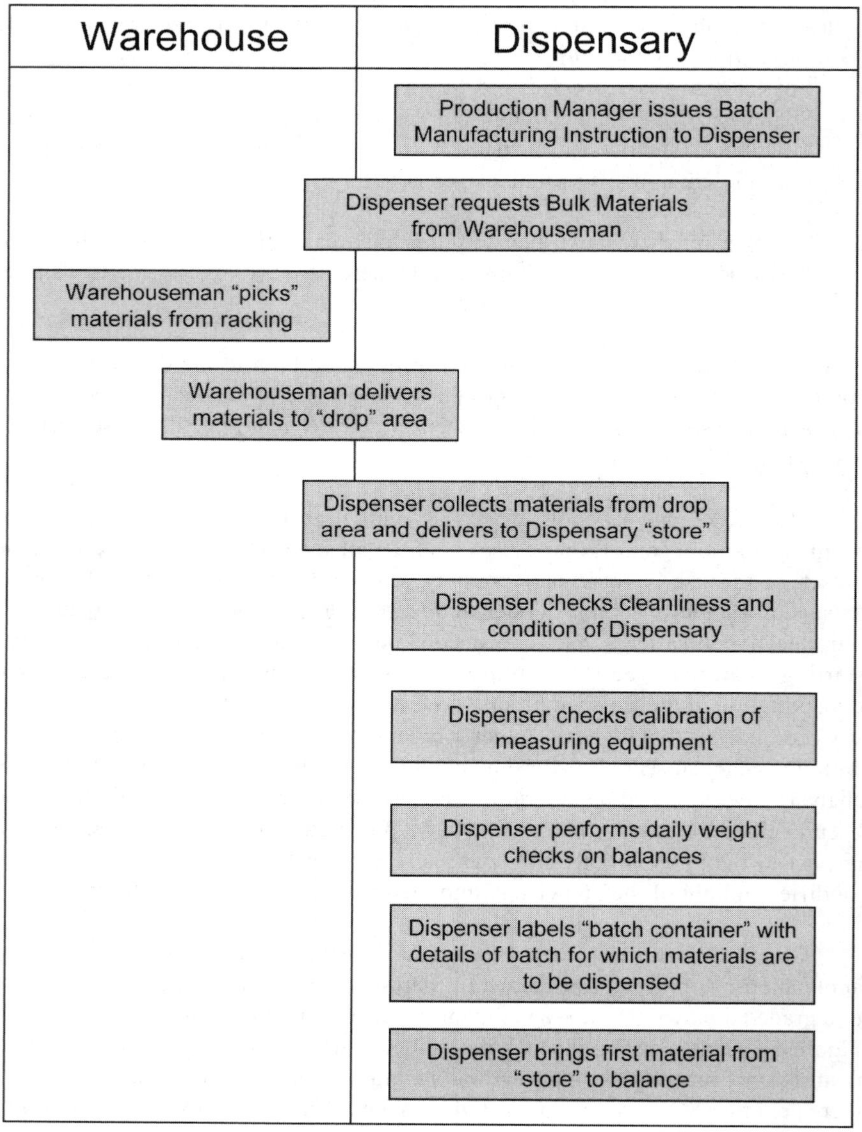

Figure 5.4 Partial Process Map Showing Movement of Material through Graded Aseptic Areas and Highlighting Supporting SOPs

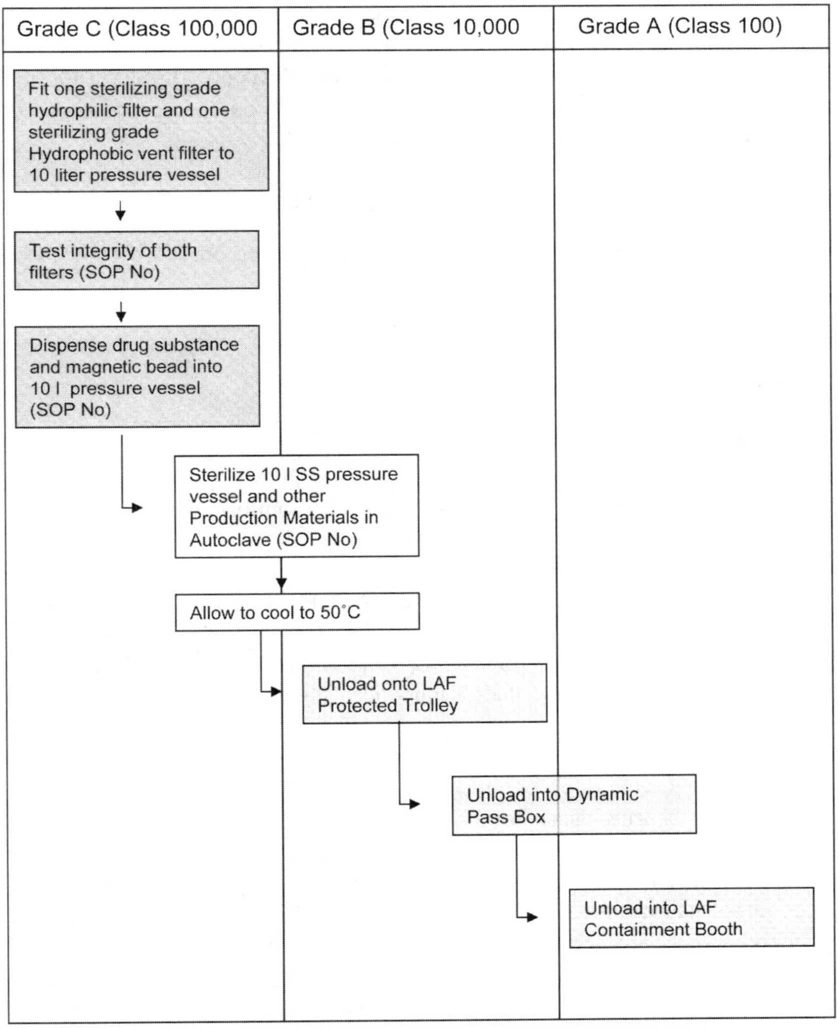

When applying formal risk analysis to pharmaceutical manufacturing projects it should be expected that the process map is independently reviewed, usually, but not necessarily by the quality organisation. It should also updated through the change control system.

Hazard Analysis Critical Control Points (HACCP)

HACCP was developed for the Pilsbury Food Corporation, and dates from 1959. It has been promoted and endorsed for food safety by various international (Food Quality and Standards Service Food and Nutrition Division, 1998; WHO, 2002) and national (FDA, 2001; Anon, 2003) bodies. FDA is extending the application of HACCP into medical device manufacture.

HACCP is an 'umbrella' tool comprising seven principles.

1. Conduct a hazard (risk) analysis.

2. Determine critical control points (CCPs).

3. Establish critical limits.

4. Establish a system to monitor each CCP.

5. Establish appropriate corrective action when monitoring indicates a CCP is not under control.

6. Establish procedures to determine the HACCP process is working correctly.

7. Establish documentation concerning all procedures and records appropriate to HACCP principles and their application.

- *Principle 1 — conduct a hazard (risk) analysis.* HACCP is not prescriptive. There is no direction for 'conducting a hazard (risk) analysis' and combined in the first of its principles is both its greatest strength and its greatest weakness.

  The strength in this lack of prescription is that it leaves the user free to employ whatever means is best for conducting the hazard analysis. Some risk analysis tools are better than others for particular applications, while some may require adaptation. The risks in some applications may be better analysed by wholly novel approaches. Its weakness lies in that it gives no guidance to the novice as to how to set about organising a risk analysis except to the extent that it requires the creation of a process map as a preliminary step.

- *Principle 2 — determine critical control points*. A critical control point is defined as a process step at which necessary action can be applied to ensure and maintain compliance with specified conditions/limits. This is not always easy to understand in the pharmaceutical context, where processes are generally 'fixed' by license conditions and adjustment may not be permitted. A critical control point which does not allow adjustment is of little value.

  An example of an existing permitted *critical control point* could be (say) the case of neomycin sulphate (active pharmaceutical ingredient)where it is well known that the actual biologically active neomycin content differs from batch to batch.

  - The risk from dispensing a fixed quantity is that there may be an unacceptable variation in neomycin content in the finished dosage form.
  - The critical control point comprises analysis on receipt. The data obtained not only facilitates accept–reject decision making, but also allows correct amounts to be dispensed according to its biological activity.

  In furtherance of its initiative *Pharmaceutical GMP's for the 21st Century — a Risk Based Approach* (FDA, 2004(a)) — the FDA sign-posts future possibilities for expanding critical control point philosophies. This part of the initiative is called *Process Analytical Technology* (PAT) (FDA, 2004(b)).

  PAT is intended to encourage companies to develop and apply better process understanding to pharmaceutical manufacture. It encourages use of on-line/at line/in-line process and product monitoring with a view to making real-time process control adjustments to ensure end product quality. For example, it describes how the familiar time-based process end-points (eg blend for 10 minutes) may not take account of physical differences among batches of starting materials which have traditionally been required only to comply with pharmacopoeial limits for chemical identity and purity.

  Among pages of repetitive self-justification, the PAT initiative (FDA, 2004(b)) recognises there may be perceived barriers to implementation of its principles (eg pre-approval inspection). It puts forward a regulatory strategy to facilitate a consistent scientific assessment of marketing applications, amendments and supplements, designated as being within its remit.

- *Principle 3 — establish critical limits*. There is nothing unusual about establishing limits for pharmaceutical processes, products, intermediates, etc. Sensible validation is, after all, based on the pre-determination of acceptance criteria (limits). However these may not always be critical limits.

Figure 5.5. 'Capable' Processes Operating within Registered Limits. (a) Two processes with central tendencies near the mid-point of the allowable band but with different 'spreads' of data. (b) Two processes with similar 'spreads of data, one operating close to the upper registered limit, the other operating close to the lower registered limit.

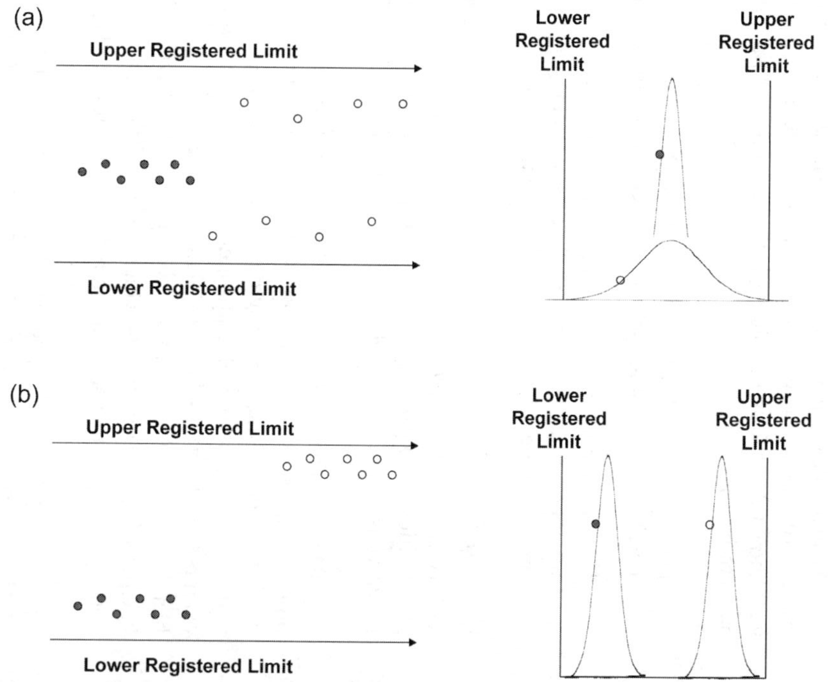

Limits and acceptance criteria in pharmaceutical manufacture may be taken from the pharmacopoeias, from existing *Standards*, from regulatory guidance, from experience of past difficulties or successes at registration or inspection, etc. Critical limits, on the other hand, should be based on process capability. Figure 5.5 illustrates four processes all of which are capable of operating within and complying with the registered limits, but each of which actually performs differently. Figure 5.5(a) shows two processes with central tendencies close to the mid-point of the registered limits — one has a tight distribution, the other a wider spread of data. Figure 5.5(b) shows two processes with tight distributions — one operates with a central tendency close to the upper registered limit, and the other with a central tendency close to the lower registered limit. What is not shown but should be self-evident is the impact of a process which has a wide spread of data but with a central tendency close to one of the registered limits — these should of course 'fail' validation and be rare occurrences.

The risk within a process is going to relate to the manner in which the data are actually distributed and how this relates to other limits. Setting *critical limits* requires good process knowledge.

- *Principle 4 — establish a system to monitor each critical control point*. Again this is not an unfamiliar concept in pharmaceutical manufacture, except HACCP focuses its monitoring requirement on the critical control points as distinct from end-product testing, where much pharmaceutical quality testing takes place.

- *Principle 5 — Establish appropriate corrective action when monitoring indicates that a CCP is not under control*. The pharmaceutical industry already has systems for addressing corrective action which have been mandatory requirements for many years and have been subject to inspection.

- *Principle 6 — Establish procedures to determine that the HACCP process is working correctly*. The concepts of self-inspection and audit of HACCP systems are no different from those applied in other areas of pharmaceutical manufacture.

- *Principle 7 — Establish documentation concerning all procedures and records appropriate to HACCP principles and their application*. Again, the pharmaceutical industry is probably the least naive of all industries in respect of requirements for formal and approved documentation, and documentation control. There is no standard approach to documenting HACCP studies.

In summary, HACCP is an 'umbrella' risk analysis tool which is non-prescriptive to identification of risks. To optimise its value, critical control points must be established. If monitoring data exceeds limits set sensibly around them, these can then be addressed. This may be a problem with registered pharmaceutical manufacturing processes, but FDA's PAT initiative points to a way ahead which may eventually become the future norm.

In many respects HACCP presents other wisdoms — implementation of corrective actions, self-inspection, audit, documentation, documentation control etc — already in place in pharmaceutical manufacture, probably to a far greater extent than they were in the food industry pre-HACCP. If a pharmaceutical company decides to implement HACCP as a risk analysis tool it should avoid duplicating existing systems, but accept that dovetailing into them may not be straightforward.

Failure Modes and Effects Analysis (FMEA)

FMEA has its origins in the US military in the 1940s, was adopted by NASA in the 1960s and by the Ford Motor Company in the 1970s. It generates action plans to prevent, detect or reduce the impact of things that might go wrong. It lends itself best to analysis of risks associated with equipment.

FMEA can be applied to test a design at an early stage in a project (design FMEA), or to improve an established process (process FMEA). The strength of FMEA is in its structured approach to anticipating how a design, a piece of equipment, or a process, can fail. It is underpinned by the answers to three questions:

- What might go wrong?
- How badly would it affect things if it did go wrong?
- What might cause it to go wrong?

FMEA hinges on the completion of a form (the FMEA form).This is just a structure applied to a thought process. Completion of the form requires the identification and evaluation of potential failures (risks, hazards, etc). These are allocated a 'score' in the form of a prioritisation number (the Risk Priority Number (RPN)).

The allotment of an RPN to potential failure modes provides a practical priority order for corrective action. After identification of each risk, the RPN is calculated by multiplying together three 'scores' determined for each risk against arbitrary scales:

- How severe is the effect of the potential failure going to be?
  - This is called the 'S' scale — the score number which could be on a scale of 1–3, or 1–5, or 1–10 gets bigger as the severity increases. Scales are essentially determined arbitrarily to suit the particular subject under analysis. Obviously scales of 1–10 multiplied by each other three times give the RPN greater discrimination (1–1,000) than scales of 1–3 (RPNs between 1–27) or 1–5 (RPNs of 1–125). Scales of 1–5 are most commonly used.
- How likely is an identified potential failure to occur?
  - This is called the 'O' scale — the score number gets bigger as the probability of occurrence increases.
- How likely is it that the potential failure will be detected?
  - This is called the 'D' scale — in contrast to the other two scales the score number gets smaller as the likelihood of detection increases.

The FMEA form is standard (see Figure 5.6). Its use is probably best illustrated through an example (Figure 5.7).

Figure 5.7 represents an element of an FMEA on that part of a steam sterilisation process in an autoclave which takes place before the cycle starts — preparing materials for sterilisation, loading them in trays or trolley racks for sterilisation and closing the sliding door of the steriliser.

- The headings show that the FMEA has been allocated a unique code number, the team leader and technical secretary are identified, and it is clear that this FMEA applies to a particular type of sterilisation process in a particular autoclave.
- The 'Process Description' shows which part of the FMEA is being documented and refers to a Process Map to clarify the study parameters.
- The Failure Mode highlighted is a simple one — the autoclave door fails to slide close.
- Under the column marked 'Effect of Failure' it is stated that the autoclave cycle cannot start unless the door is closed and that this, where limits are placed on the time between filling the product and sterilising it, may result in

Figure 5.6 The FMEA Form (annotated)

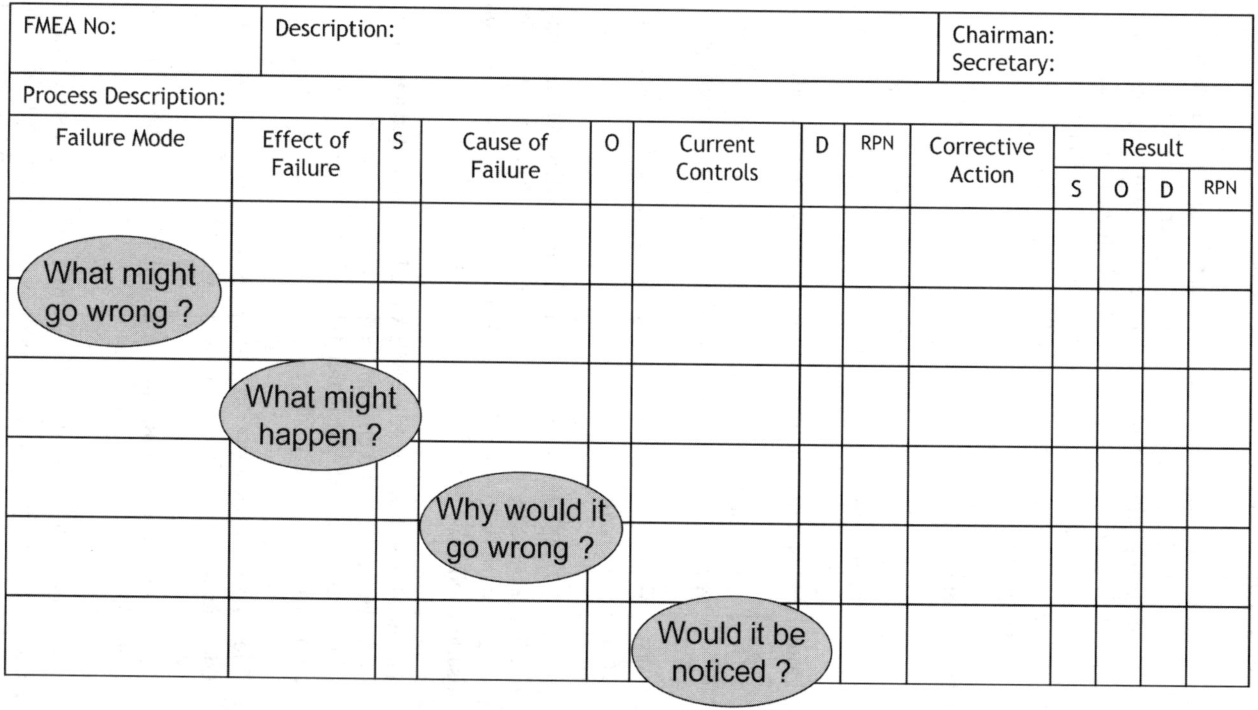
FMEA No:
Description:
Chairman:
Secretary:
Process Description:
Failure Mode
Effect of Failure
S
Cause of Failure
O
Current Controls
D
RPN
Corrective Action
Result
S
O
D
RPN
What might go wrong ?
What might happen ?
Why would it go wrong ?
Would it be noticed ?

Figure 5.7 Illustration of Use of the FMEA Form

FMEA No:nnn06	Description: FMEA on Sterilization of Liquid Loads in Autoclave 1234							Chairman: Abdul Bates Secretary: C Doughnut				
Process Description: Process stages prior to beginning sterilization cycle (see Process Map xxx, Attached)												
Failure Mode	Effect of Failure	S	Cause of Failure	O	Current Controls	D	RPN	Corrective Action	Result			
									S	O	D	RPN
4. Autoclave door fails to close	Cycle cannot start, Product may require rejection or re-work	5	4.1 Power supply 4.1.1 Electricity supply failure 4.1.2 Fuse blown	 1 1	Further process does not start	1	**5**					
			4.2 Motor failure	2	Further process does not start	1	**10**					
			4.3 Tracks obstructed	4	Further process does not start	1	**20**					

the product having to be rejected or re-worked. The severity on the 'S' scale is given a score of 5 on a 1–5 scale.

- Three broad reasons why the door may not close are listed under 'Cause of Failure' — power supply, motor failure and tracks obstructed. Power supply as a cause of failure has been sub-divided into electricity supply failure and fuse blown. Other potential causes of failure could have been sub-divided too, but for this illustration further depth is unnecessary.

- Against cause of failure, the 'O' (occurrence) scale has been completed. In this case electrical failures have been allocated a score of 1 — very unlikely, but this is not, of course, the case in all locations. Motor failure has been allocated a score of 2 — if sub-divisions were undertaken there could be different scores for various parts of the motor. Track obstruction has been given a score of 4, quite probable, as anyone who has experience of sliding doors might expect.

- Under current controls the FMEA form states 'further process does not start' — in other words it would be immediately evident that the door had not closed, therefore the probability of detection is very high and a low number is allocated on the 'D' scale.

- Three RPNs have been calculated from these individual 'S', 'O' and 'D' scales — the highest, with an RPN of 20, points to the possibility of the track obstruction. If any action were to arise from this FMEA, this would be the topic given highest priority. In fact, using scales of 1–5, RPNs of 20 are usually fairly low on the overall priority list, with RPNs higher than 60 usually those requiring immediate attention.

- The columns on the right-hand side of the form record corrective actions completed, and allow for calculation of a new RPN.

FMEA is highly structured but relies heavily relies on the initial identification of risks (failure modes) — usually best done by taking linear segments of processes, and at each stage asking the question, 'what might go wrong?'. This is wholly dependent upon a good process map, or key steps may be missed.

Using the operation of an autoclave as an example, steam injection to bring the chamber and load to the correct operating temperature could again be taken as an instance.

- What might go wrong?

- Temperature overshoot.
- Operating temperature not reached or slow progress to operating temperature.

- What might cause these things to go wrong?
  - Steam valve stuck 'open' or partially open (temperature overshoot).
  - Steam valve stuck partially open.

There may also be other reasons and it is important that the team has the expertise to identify them. Thereafter it is within the team's proficiency to allocate scores on the 'S', 'O' and 'D' scales.

FMEA documentation lies in completion of the form. As risk management becomes a recognised feature of pharmaceutical manufacture and FMEA is used, it should be anticipated that the completed form will become an inspectable document. In other words, no company using FMEA can afford to have identified a risk with a high RPN and then fail to take corrective action. In risk management circles, improved and/or increased detection methods are frowned upon because they do not get to the root cause of the risk. In pharmaceutical manufacture, however, it could be that improved and/or increased detection has to be a short-term solution, at least until any more fundamental changes are given the regulatory 'go-ahead'.

In summary, FMEA is a means of obtaining a prioritised action plan. It is systematic and has a standard form of documentation.

PRACTICALITIES OF RISK MANAGEMENT IN PHARMACEUTICAL MANUFACTURE

Application of 'one size fits all' concepts to risk management, for instance using unmodified standard risk analysis tools such as HACCP, FMEA, etc, may not be best suited to all the particularities of pharmaceutical manufacture. It is important that the principles should be well understood because their value has stood the test of time in other industries. On the other hand, the pharmaceutical industry should not be afraid to modify them for specific needs or even use other approaches if necessary.

Various potential approaches to particular situations are described below.

Aseptic Manufacture

With 21st century technology, it is generally accepted that the greatest risk to product sterility arising in aseptic manufacture is the involvement of people. Personnel activities in areas where the sterile dosage form, or sterilised product-contact components and/or pieces of equipment are protected, have been identified as risk areas and should be included in process simulation (FDA, 2004(c)).

Asepsis means the complete avoidance of contamination, and aseptic manufacture of sterile drug products provides an example of how effective this can be. Actual instances of patient harm arising from administration of supposedly sterile (but actually non-sterile) dosage forms manufactured aseptically have been very few. However, the application of standard risk analysis tools like FMEA, HACCP to aseptic manufacture typically comes up with the same conclusions — most likely cause of potential contamination — personnel; severity — high; occurrence — low; detection — low. This points to removal of people, perhaps through isolation technology, as the best means of managing risk in aseptic manufacture. However, this a radical and expensive solution where process improvement may be the only genuinely practical objective.

This may require aseptic manufacture (based on its own peculiarities and particularities) to take an alternative approach to risk analysis. For instance, some manipulations required by personnel in aseptic manufacture create greater risks of contamination than others. It is worth considering why this should be.

- *Proximity*. FDA (2004(c)) emphasises that sterile materials should only be handled using sterile instruments and that personnel should have no direct contact by any part of their gown or gloves with sterile products, containers, closures or critical surfaces. Aside from the evident impracticality of this provision to all circumstances, (for example, assembly of sterile filling pumps) it is clear that the closer personnel are to sterile materials the more likely they may inadvertently contaminate those materials. The containment of personnel in sterile garments, gloves, etc can never be perfect, nor can their aseptic disciplines. Some shedding of microorganisms from their skin, which then escapes into the environment, will occur. Protective unidirectional air flow is intended to blow airborne microorganisms away from protected items. Turbulent air flow is intended to prevent microorganisms settling on protected items. As personnel come closer to protected items or introduce instruments or parts of their bodies into areas close to protected items, the risk of disrupting protective air flow close to the protected items increases. The probability of microorganisms from personnel contaminating protected items increases as they come closer to protected items.

It must be emphasised that proximity is not solely a matter of closeness to the dosage form itself, but to any other item of equipment or container that comes into contact with the sterile dosage form. Microorganisms shed from personnel onto rubber vial stoppers, while unloading them from an autoclave or loading them into a hopper, have the same probability of causing non-sterility, as microorganisms shed into the Grade A (Class 100) area around point of fill.

- *Duration*. In natural environments people shed microorganisms all the time, as they breathe, move and blink. Some people shed more than others. Garments provided to assist in their work in aseptic areas are intended to contain these shed microorganisms. The probability of these garment-barriers being overcome is a function of number of microorganisms shed — this in turn may be a function of the rate at which microorganisms are shed, and of the elapsed time during which they are shed. Aseptic manipulations in the region of protected items which take longer are more likely to result in contamination than operations which take shorter periods.

  Personnel intervention into aseptic areas for collection of samples for fill volume or fill weight checks is generally a quick aseptic manipulation, often taking as little as five seconds. Aseptic assembly of pre-sterilised filling pumps (pistons into cylinders, pumps onto manifolds, manifolds onto machines) on the other hand may take anything from 30 minutes to more than an hour. Modern aseptic filling equipment can be purchased with robotic sampling, weighing and recording devices. Filling machines with integral clean-in-place, sterilise-in-place systems covering filling pumps and needles are also available — regrettably they are not ideally suited to efficiency in manufacture of different container sizes, fill volumes, etc.

- *Complexity*. Some tasks which may appear simple 'on paper' can be quite complex when put into practice.

  For instance, connecting two flexible hoses using a tri-clover connection appears quite easy 'on paper'. In practice it requires some considerable dexterity. The author has heard experienced aseptic area personnel jokingly explaining that the 'tri' in tri-clover means it requires three hands to make the connection. Assembly of filling pumps is certainly intricate, requiring tools such as Allen keys, spanners, and possibly screw drivers. The greater the complexity in an aseptic manipulation the greater the likelihood that aseptic disciplines will be sacrificed in favour of getting the job done.

The probability of success or failure in aseptic manufacture largely lies in understanding the minutiae of these three risk factors — proximity, duration,

Table 5.1 Suggested Scores on a 1–5 Scale for Mitigation Factors for an A-RPN Risk Analysis Tool for Aseptic Manufacture

Score	5	4	3	2	1
This suggested system assumes that:	Personnel trained only in principles of asepsis				
1 Aseptic processes are documented in detail		Personnel trained in specific aseptic manipulation			
2 Training is given by competent personnel			Personnel trained and assessed by simulation		
3 Unidirectional air-flow protection has been evaluated in operation by smoke studies				Training assessed by simulation and operation protected by Unidirectional air-flow	
4 Barrier or isolation technology has been properly validated and controlled.					Training assessed by simulation and operation protected by isolation or barrier technology

complexity. Scores may be applied to compute a variation on the RPN of FMEA. An Aseptic Risk Priority Number (A-RPN) can be calculated by multiplying Proximity (P), Duration (D) and Complexity (C) scores. Sensible application of scores can only, however, be done from close observation of how personnel tackle their day-to-day roles. 'Better' ways of aseptic manipulation can only emerge from knowing exactly how difficult these jobs may be. In other words, a risk management tool can be applied but it can only contribute to process improvement if underpinned by process understanding.

The A-RPN derived from $P \times D \times C$ does not take into account any mitigating factors operating within the aseptic manufacturing process in the way a detection scale applies in standard FMEA. A further score can be given to the effort a company has put into reducing the probability of contamination actually occurring. This scale would, of course, operate inversely to the P, D, C scales, with low numbers applied to the highest level of protection afforded (eg isolation technology) and high numbers to the lowest (eg personnel trained but not assessed in their ability to perform the manipulation aseptically). A suggested approach to developing a mitigating scale is given in Table 5.1.

Risk Benefit Decisions Relating to Contamination Control

Perfection is impossible, total purity unobtainable. There must always be some risk of contamination arising. In pharmaceutical quality and GMP regulation there is an expectation, sometimes explicit, but more often implied, that 'substances' and processes should be sampled, tested, or monitored in some way to 'track' contaminants, or to allow 'pass–fail' decision-making with regard to contamination, or to stimulate corrective or preventive actions, process improvements, etc.

The activities involved in sampling (or monitoring) may sometimes introduce a risk of contamination into a process in aseptic manufacture which might not otherwise have arisen. In such cases 'testing' may not exemplify a conscientious approach to product quality. Examples include application of the *Test for Sterility* to incoming sterile materials, and microbiological monitoring of/in critical (Class 100, Grade A) aseptic manufacturing areas.

- *The Test for Sterility*. Introduced in the *British Pharmacopoeia* in 1932 and in the *United States Pharmacopeia* in 1938, the *Test for Sterility* was initially presented as a relatively uncomplicated method applicable to finding out if items tested were sterile or non-sterile. Sampling schemes first appeared in the *UK Therapeutic Substances Regulations* in 1952, in the *United States Pharmacopeia* in 1955 and in the *British Pharmacopoeia* as late as 1968. These sampling schemes and sampling statistics came under heavy criticism

Table 5.2 Risk Benefit Analysis (Partial) for Applying the Test for Sterility to Sterile Starting Materials

Risk Benefit Analysis: *Test for Sterility on Sterile Starting Material xxx*			
Participating: Abdul Bates/Catherine Doughnut/Edith Fang		Date Performed: *nn nn nnnn*	
Benefit or Perceived Benefit	Risk or Perceived Risk	Comments	Conclusions
Provides assurance that each batch is sterile		Statistical limitations are such that the *Test* does not provide assurance of sterility	Benefit value — Low
	Batch may become contaminated during sampling	If sample is also contaminated this might be disclosed by the *Test* On the other hand, it might not be disclosed and present a risk to the patient. Provision of sampling training and protection should minimise this risk	Risk value — High (possible impact on patient)
	Sample may become contaminated in collection or testing	False positive result although conservative initiates needless effort into investigation and may damage relations with supplier. Provision of isolator for testing should minimise this risk	Risk value — Medium (possible impact on cost and business relationships)
Conclusions	The risks significantly outweigh the benefits. Annex I to this Analysis lists other indicators of sterility for sterile Starting Material xxx.		
Recommended Action and Responsibilities for Action			

even from before they were formally incorporated in the compendia (Knudsen, 1949; Maxwell Bryce, 1956; Brown and Gilbert, 1977). Quite simply they have no statistical power to qualify a batch of items for sterility against the standard accepted for sterility of a probability of not more 10^{-6} for an item being non-sterile.

The Test for Sterility, however, continues to be applied in many laboratories, to sterile starting materials supplied in bulk. Is the value obtained from applying this test outweighed by the risks arising from it? This question can be answered by performing and documenting a Risk Benefit Analysis (see Table 5.2).

The benefits and risks are identified in Table 5.2. The analysis form allows comments against each benefit or risk. These should fairly reflect the views of the participants. The analysis form requires some summary conclusion against each benefit or risk (high — patient benefit or risk should always be high — medium or low) and a final set of conclusions weighing up the whole analysis. In the example (Table 5.2) of performing the *Test for Sterility* on sterile starting materials, the benefit relative to providing assurance of sterility of the batch is low, but the attendant risks of contaminating the product are evaluated as medium to high. Thus the final conclusion is that the risks of performing this *Test* outweigh the benefits. In such cases the conclusion should indicate alternative means of assuring sterility. These may include supplier audit and responsiveness to recommendations for process improvement. The supplier may have an exemplary inspection history of pass results in previous *Tests for Sterility* on the starting material or on the finished product (although the poor statistical power of the *Test for Sterility* is well known, the perception is that a good history of *Sterility Test* results is somehow valuable). In effect this particular risk analysis points the way to 'parametric receipt'.

- *Microbiological Monitoring in Critical Areas*. The FDA *Guidance for Industry: Sterile Drug Products Produced by Aseptic Processing* (FDA, 2004(c)) emphasises the importance of critical surfaces coming into contact with product remaining sterile throughout manufacture, and that these locations (because they pose the greatest microbiological risk to the product) should be included as part of the environmental monitoring programme. FDA continues (FDA, 2004(c)) 'Critical surface sampling should be performed at the conclusion of the aseptic processing operation to avoid direct contact with sterile surfaces during processing'. This exemplifies a risk-based conclusion reached most probably intuitively, or through common sense. FDA does not elaborate a process of reaching this conclusion, but the principle is there to be applied with appropriate justification and documentation in other circumstances.

Table 5.3 Risk Benefit Analysis (Partial) for Microbiologically Monitoring Surfaces of Sterile Product Contact Equipment used in Aseptic Manufacture

Risk Benefit Analysis: *Microbiological Monitoring of Sterile Product Contact Surfaces During Manufacture*			
Participating: Ghengis Harris/Idris Jiang/Ken Lee		Date Performed: *nn nn nnnn*	
Benefit or Perceived Benefit	Risk or Perceived Risk	Comments	Conclusions
Provides assurance that product contact equipment is sterile		Methodology insensitive to detecting non-sterility, media growth support tested only at greater than 10 cfu; recovery from equipment unlikely to be 100%	Benefit value — Low
Provides assurance that product is not likely to be contaminated through contact with contaminated equipment		Should equipment be contaminated at detectable levels it is possible that there may be enough present to contaminate product.	Benefit value — High
	Product may become contaminated during monitoring of equipment	May not be detected by insensitive end product *Test for Sterility*. Provision of sampling training and protection should minimise this risk.	Risk value — High (possible impact on patient)
	Sample taken for monitoring may become contaminated in collection or testing	False positive result although conservative initiates needless effort into investigation and could lead to unnecessary batch rejection.	Risk value — Medium (possible impact on cost and business relationships)
Conclusions	The risks marginally outweigh the benefits. There are only periodic simulation data to indirectly support maintenance of sterility of sterile product contact equipment surfaces during operations.		
Recommended Action and Responsibilities for Action	Monitor sterile product contact surfaces only at the conclusion of a batch — Microbiological Environmental Quality Group to implement in conjunction with Production.		

Table 5.3 illustrates how FDA's conclusion could possibly be justified and documented.

- The first perceived benefit is that monitoring could provide assurance that the product contact surface was actually sterile. In fact this is not possible.
  - The standard media growth support test is done against a 'low' inoculum of usually 10–100 cfu, in other words, the media have not been not proven as indicators of sterility.
  - It is unlikely that either swabs or contact plates (the probable techniques for surface monitoring) are 100% efficient at removing microorganisms from surfaces and subsequently transferring them to media.
- The second perceived benefit is that monitoring could indicate if product could be contaminated through contact with contaminated equipment. Accepting the limitations identified above, it would appear that monitoring is probably only going to recover microorganisms when significant numbers are present on the surfaces, and that in such numbers there is a reasonable probability that some could be transferred to and contaminate the product. This would certainly appear to be well worth knowing.
- On the risk side, the act of monitoring could itself lead to contamination of both equipment and product which could have remained sterile had monitoring had not taken place. There is really no way of knowing if this could have happened, it would be naive to expect the *Test for Sterility* on end-product samples to necessarily disclose contamination caused by monitoring. The risk is to the patient — administration of a non-sterile dosage could result in septicaemia, endotoxic shock, even death.
- The second risk area is contamination of the monitoring sample leading to a false positive result, investigation, inappropriate or unnecessary actions, possible unnecessary batch rejection.

So, in the example given in Table 5.3 the balance between risk and benefit is not quite as clear as in the example given in Table 5.2 with respect to the *Test for Sterility* on sterile starting materials. On balance, the conclusion drawn on monitoring the surfaces of sterilised product-contact equipment has been that the benefit of knowing if it could have become contaminated in use to the extent that it risked transfer of contaminants to the product, is marginally outweighed by the risk that the act monitoring could itself contaminate the product. The final recommendation is that this type of monitoring is worth doing, but only at a time when there is no risk of it contaminating the product, namely after conclusion of manufacture.

CONCLUSIONS

No manufacture is without risk of defect in some items: pharmaceutical manufacture cannot be excluded from this generalisation.

Although in the past risk has not been emphasised in relation to pharmaceutical manufacture, it has existed and has been addressed, perhaps intuitively, or perhaps through formal risk analysis by manufacturing equipment suppliers. The impact of recent regulatory changes is doubtlessly that risk management in pharmaceutical manufacture will in the future be conducted more systematically and more formally. There is much to be learned from the tools used in other industries (which perhaps may, in their turn, have something to learn from the pharmaceutical industry about documentation and formal recording of processes). However these risk analysis tools are unlikely to be amenable to pharmaceutical manufacture without sensible modification to meet their new purposes.

REFERENCES

Anon (1986) *The Penguin Pocket English Dictionary*. London: Penguin Books in association with Longman Group Ltd.

Anon (2002) *The Rules Governing Medicinal Products in the European Community. Volume IV. Good Manufacturing Practice for Medicinal Products.*

Anon (2003) *Improving Service Delivery — the Food Standards Agency — a Report by the Comptroller and Auditor General HC 524 Session 2002–2003: 28 March 2003.* London: The Stationery Office.

Anon (2006) Review of FDA Warning Letters issued in 2005. *The Pharmaceutical and Healthcare Sciences Society (PHSS) GMP Update, March 2006.* Swindon, UK: The Parenteral Society.

Brown, M.R.W. and Gilbert, P. (1977) Increasing the probability of sterility of medicinal products. *Journal of Pharmacy and Pharmacology* **29**, 517–523.

Code of Federal Regulations Part 123 *Fish and Fishery Products.*

Code of Federal Regulations Part 211 *Current Good Manufacturing Practices for Finished Pharmaceuticals.*

FDA (2001) *Fish and Fisheries Products Hazards and Controls Guidance.* US Food and Drug Administration, Center for Food Safety and Applied Nutrition, June 2001.

FDA (2004(a)) *Pharmaceutical GMP's for the 21st Century — a Risk Based Approach Final Report.* Department of Health and Human Services, US Food and Drug Administration, September 2004.

FDA (2004(b)) *Guidance for Industry: PAT — a Framework for Innovative Pharmaceutical Manufacturing and Quality Assurance* Department of Health and Human Services, US Food and Drug Administration, CDER, CVM, ORA, August 2003.

FDA (2004(c)) *Guidance for Industry: Sterile Drug Products Produced by Aseptic Processing — Current Good Manufacturing Practice* Department of Health and Human Services, US Food and Drug Administration, CDER, CBER, ORA, September 2004.

Food Quality and Standards Service Food and Nutrition Division (1998) *Food Quality and Safety Systems — A training manual on food hygiene and the hazard analysis and critical control point system* Rome: Food and Agricultural Organization of the United Nations.

ICH (2005) *Harmonised Tripartite Guideline- Quality Risk Management Q9* Step 4 version November 2005. Geneva, Switzerland: International Conference on Harmonisation of Technical Requirements for Registration of Pharmaceuticals for Human Use.

Knudsen, L.F. (1949) Sample size of Parenteral solutions for sterility testing. *Journal of the American Pharmaceutical Association* **38**, 332–337.

Maxwell Bryce, D. (1956) Tests for the sterility of pharmaceutical preparations. *Journal of Pharmacy and Pharmacology* **8**, 561–572.

World Health Organization (2002) *WHO Global Strategy for Food Safety: safer food for better health*. Geneva, Switzerland: WHO

6

BACTERIA RETENTIVE FILTRATION

Simon Cole

INTRODUCTION

Filtration is employed in practically every pharmaceutical manufacturing process today, either to improve the quality attributes of services, raw materials and intermediates or to assure the quality of the end product and the environment in which it is formulated into its finished dosage form. Microporous filter membranes with a removal rating (porosity) of 0.2µm (or 0.22µm) assigned by their manufacturers are routinely used today for the sterilisation of pharmaceutical drug products. They were introduced over 30 years ago to replace the 0.45µm grade previously used. Evidence had emerged of the existence of bacteria smaller than those used formerly to define the performance of sterilising filters and a new standard was established. In certain industrial applications manufacturers have chosen to use still finer filters, grades classified as 0.1µm or even finer — 0.04µm and 0.02µm, that is, 40nm and 20nm, respectively — for the removal of specifically identified threats to their processes, such as deformable mycoplasma-type bacteria and viruses. Despite their finer porosity and ability to retain much smaller particles, these filters have neither replaced nor changed the regulatory definition applied to sterilising grade filters for use in aseptic production processes.

In the early days of sterile membrane filtration at industrial scale, procedures involved installing a relatively-brittle sterilising grade membrane in a heavy filter

Figure 6.1 Examples of capsule and stainless steel filter assemblies

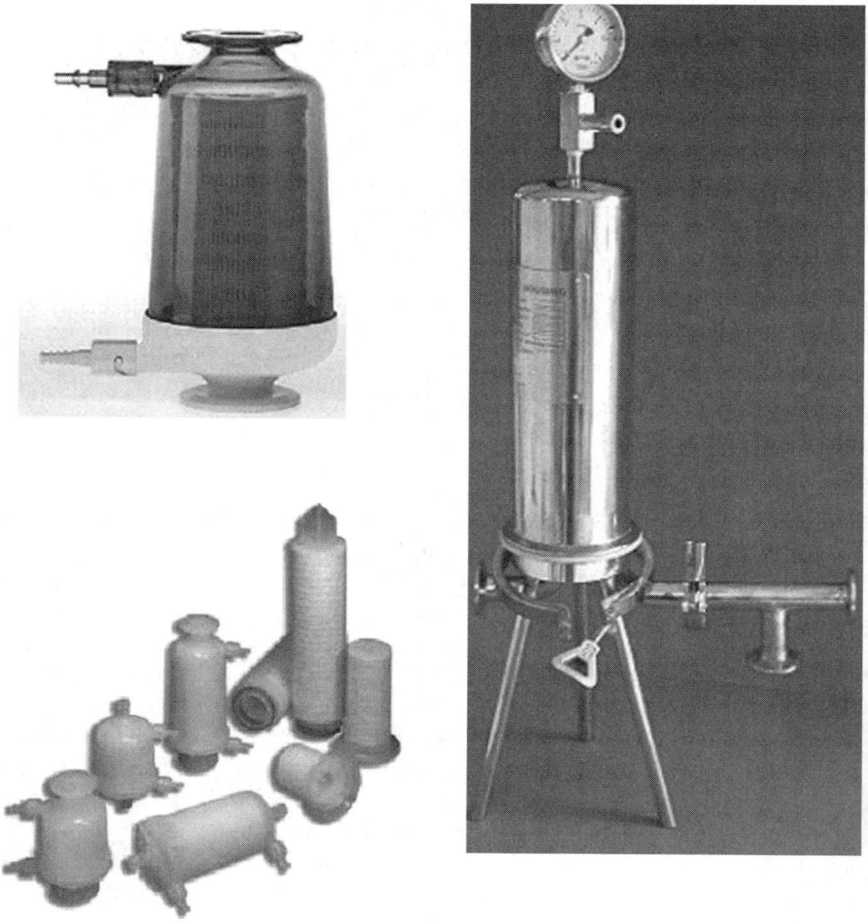

disc holder that was hard to seal effectively. Membrane flow rate capability was poor and to achieve the necessary throughput for industrial-scale applications a multitude of such holders was employed in a manifold arrangement. Although still used surprisingly widely today, the risks that they present to aseptic processes have led to their replacement in more highly-regulated manufacturing environments.

The introduction of pleated and stacked-disk filter cartridges marked a major change in filtration technology, offering a means to enclose all the surface area of filtration membrane required for a specific duty in a single container, the filter housing, using single- or multi-element filter cartridges. Filter suppliers next sought to minimise the necessary size of these housings by improving the materials and methods of manufacture, increasing the flow capacity and retention capability of the membrane and other filter components, while optimising the fabrication techniques and cartridge configuration.

Replaceable filter cartridges in steel housings are not suited to every situation for reasons of practicality, regulatory compliance, GMP or economics. Suppliers developed ranges of self-contained, fully-disposable filter *capsules*, comprising a filter element integrated in a light-weight plastic outer casing. These have been favoured by a bio-pharmaceutical industry manufacturing finished products from materials of biological origin, where filtration practices have been re-examined. The need to assure product purity and avoid batch-to-batch carry-over stimulated interest in the use of fully disposable equipment and systems, particularly for use with biologically-sourced materials whose precise composition is not entirely consistent. Using disposable materials eliminates the need to clean a lot of processing equipment and hence avoids the costs of both cleaning and the validation of cleaning. Many of the process risks associated with cross-contamination and product carry-over between batches are also reduced or eliminated.

Sterilising filters incorporated in disposable plastic capsules cannot generally be sterilised by in-line steaming for safety reasons — they must be sterilised before use by autoclaving or another treatment (irradiation, ethylene oxide or vapourised hydrogen peroxide, for example). This does not present a handicap to their use as part of fully disposable processing systems, since these latter systems must themselves be similarly pre-sterilised. Enhanced aseptic handling technologies represented by flexible film barrier isolators have brought improved aseptic manipulations to the sterile manufacturing process arena, reducing the advantages previously offered by the steam-in-place capability of stainless steel filter systems.

Validation standards and requirements for sterilising filtration systems have changed substantially during this 0.2μm era (Cole, 1995). In the early days filter manufacturers could be separated into those who validated and those who did not. The selection of a filter supplier offering printed validation documents reduced the further validation effort required by the user for a sterilising filtration process. The landscape changed in 1987 with the publication of the United States Food & Drug Administration (FDA) *Guideline on Sterile Drug Products produced by Aseptic Processing* (FDA, 1987). Thus supplier validation guides have become helpful as an aid to filter selection by the user but of little practical benefit for validation of the filters in the actual drug manufacturing process. FDA redirected the focus to process-specific validation, ensuring that the sterilising grade filter was capable of performing to the required standard under the actual (worst case) conditions created by the combined effects of the process and the product solution. At the same time other properties of sterilising grade filters, hitherto less rigorously considered, came to be regarded with greater importance. Attention was placed on the possibility of removing components from the drug solution by adsorption, or adding to the solution's components by extracting substances from the filter's materials of construction.

Thermal sterilisation of equipment (including filters themselves) using saturated steam in an autoclave is a documentable, physical, but non-measurable phenomenon with a quantitative outcome — the Sterility Assurance Level (SAL) assigned to the treated items. By contrast, and despite many efforts, the same principles cannot be applied to sterilising filtration — a procedure more akin to magic or some sort of *black-box technology*, in which fluid goes in one side of the filter and emerges from the other without effecting any easily-apparent transformation. The inability to assign a numerical value to the statistical probability of non-sterility of the filtrate (an SAL), associated with a background of marketing humbug promulgated by competing filter suppliers has created an atmosphere of suspicion, from which a comprehensively-defined process of sterilising filtration has yet to emerge. Sterilising filtration systems can therefore expect to enjoy the intense scrutiny of regulatory authorities for some time to come.

One defining characteristic of a sterilising grade filter is that it must be possible to determine the specification and performance capability of the entire filtration assembly using a means of non-destructive integrity test. Tests to directly determine the bacterial retention capability of a filter provide absolute proof of sterilising performance, but the tests are by their nature destructive — the filter cannot be used afterwards in a pharmaceutical process. Suitable procedures are validated by filter manufacturers to enable reliable non-destructive testing of sterilising membranes incorporated into disposable cartridge and capsule designs. These methods are sensitive to membrane defects and deviations from specification, they will also identify filters damaged by handling procedures. For example, steam sterilisation of filters may account for up to 80% of all genuine failures of filter integrity (as opposed to false test failures occurring through some incorrect application of the test procedure). For disposable filter cartridges installed in steel housings, the integrity of the whole filter assembly also depends on the condition of the housing seal faces, gaskets and connections.

Available methods for testing filter integrity have not changed greatly in concept over the last 30 years, with the possible exception of the introduction of the water intrusion test of hydrophobic filters for gas applications. There have been many improvements to the ways in which the different test procedures can be performed, with automation offering better accuracy and an escape from dependence on the technique or interpretation of an operator. Claims of improved test sensitivity are debatable however, since the underlying principles of test methods are unchanged. Higher levels of both accuracy and sensitivity have been achieved for measurement of the critical test parameters, which also exposes the test result to a greater degree of signal noise or interference from environmental and machine-inspired variations (temperature, humidity, electrical fluctuations, etc).

This chapter discusses current perceptions and trends in these areas of sterile filtration and examines the background and implications of various filtration practices in terms of their impact on filter validation and operation. Recent developments in disposable materials have had profound effects on some aspects of aseptic processing; these are also reviewed with respect to filtration.

The ability to remove and retain contaminating bacteria from a drug solution remains the single most important attribute of a sterilising grade filter, and one that has both attracted informed interest and spawned speculative nonsense in equal measure. Doubtless this chapter will be seen to contribute to one or other side of the debate, according to the reader's point of view.

STERILE, IN PRINCIPLE

Sterility has been commonly defined as the complete absence of living things, excluding viruses. The latter qualification serves to avoid a semantic debate about the term living when applied to viruses, whose exclusion is generally accepted by those involved in the practical aspects of operating aseptic processes. Thus, by extrapolation, *sterilising filtration* is the complete removal by filtration of all living organisms, excluding viruses. Those involved in both the biopharmaceutical industry and its regulation often refer to the use of separate virus removal steps that are neither claimed nor validated to be *sterilising*, nor, conversely, are sterilising filtration operations attributed a viral reduction ability. Nonetheless, putting aside considerations of viral survival in the majority of drug formulations, regulations for good manufacturing practice (GMP) require validated virus removal, reduction or inactivation steps as part of a process where viral contamination is a defined risk.

Perhaps it should be added that aseptic manufacture has, despite its detractors (Wällhausser, 1979), been apparently successful for many years. Contaminated medication is not causing widespread patient harm and pharmaceutical companies are not principally preoccupied in defending themselves against legal action taken by regulators and welfare groups for selling products containing viruses. We must suppose this is because the viruses with which we should be concerned are typically quite fragile, and capable of surviving outside their natural hosts for only a very short time, relative to the duration of the manufacturing process and storage time of most pharmaceutical products. In addition, the chemical nature of the majority of drug formulations presents a hostile environment in which viruses (and many bacteria) are incapable of survival.

A sterilising filter relies on the sieving action of a microporous membrane to retain bacteria and other fine particles contaminating the solution passing through it (Meltzer, 1995). This sieving mechanism is accomplished by a combination of physical (or direct) interception of the particles, which are too large to penetrate the pore structure of the membrane (Osumi et al, 1991), and other indirect mechanisms, such as inertial impaction and charge-mediated adsorption. Particle retention is also affected by particle shape and type, as shown by the ability of a validated membrane to retain *Brevundimonas diminuta* (rods, $0.3 \times 0.6\mu m$) but allow passage of latex beads (spheres, $0.48\mu m$). To the user, all these retention mechanisms combine to provide a particular removal capability for the filter under the given operating conditions.

A more complete definition of *sterilising filtration* should be 'the physical retention by size exclusion mechanisms of all detectably-viable, independently self-replicating organisms'. This wording removes questions about viruses and avoids debate about the importance of adsorption or electrostatic charge attraction among the mechanisms of so-called absolute microbial retention. Resisting use of the word *micro-organisms* also avoids the pedantry of some pundits who have suggested (Sharp, 1995) that such a definition would not otherwise exclude, *inter alia*, goldfish and other aquatic creatures.

A feature of defining sterility in pharmaceutical products is the uncertainty inherent in its determination, echoing Heisenberg's Uncertainty Principle in quantum physics. Samples may be taken from a series of small dose final containers such as vials or ampoules and subjected to microbiological assays. A sterile outcome indicates that the units tested were either indeed sterile or contained micro-organisms not detectable in the assay, or were unable to reproduce (temporarily or permanently). However, this procedure provides no data about the remaining untested containers; statistical analysis is used to infer conclusions about the probability of sterility for the remainder of the product batch. The level of contamination in the batch is presumed to be (and we hope it is) so low as to mathematically exclude the possibility of one or more microorganism in each container. A non-sterile result confirms the presence of micro-organisms, but may not distinguish between those present in the samples tested and those introduced by contamination during the assay procedure. Similar problems exist when sampling bulk volume of a product subjected to a sterilising procedure; however we also run the risk of introducing contamination into the bulk solution during the sampling procedure.

The background to sterilising filtration in the pharmaceutical industry is adequately covered in the 1998 publication of Technical Report No. 26 *Sterilizing Filtration of Liquids* (PDA, 1998). The report also discusses filtration mechanisms and characterisation of filters as an aid to selection. Passing reference is made to important properties of sterilising filters, such as the possibility of extracting substances (extractables) from the filter materials into the drug solution. When considering validation of sterilising filters, this report exclusively discusses the principle of validating the bacterial retention capability of filters. Although of overriding importance, this one characteristic alone is not sufficient to admit a so-called sterilising grade filter for use in an aseptic manufacturing process. The second major section of the report concerns testing the integrity of sterilising filters, with discussion of the theory behind the test mechanisms.

Validating sterilising grade filters in new processes is, like all other validation activities associated with new process, a prospective or forward-looking activity (Levchuk, 1994). A planned validation exercise provides protocols with predetermined evaluation methods and, most importantly, predetermined acceptance criteria against which the results must be assessed. Much of the validation effort may be undertaken simultaneously with the design of the manufacturing process, and scaled up from the development facility, or during production of the process validation batches — that is to say, concurrently, prior to regulatory approval for the total manufacturing operation.

A greater challenge is the application of current standards of validation practice retrospectively to established pharmaceutical manufacturing processes that may have yielded apparently good results for many years. Not only are there the enormous economic implications of such an undertaking, but also the risk that some aspect of the process may not meet (or be incapable of meeting) the requirements of current GMP. For example, a number of long-established liquid pharmaceutical products contain low levels of preservative ingredients to inhibit the proliferation of contaminating bacteria, particularly once the product has been opened by the user. If the preservative is effective, then under normal circumstances the failure to remove every last bacterium in a sterilising filtration may have no adverse implications, since those remaining in the solution will be killed or inhibited from growth. In this way such a risk might go undetected for the life of the product. But if we now introduce current procedures for process-specific validation of sterilising filtration, the chosen filter must be challenged with a very large number of test bacteria under conditions that simulate the actual process. These must include the use of the actual pharmaceutical solution, but with any bactericidal component omitted and if possible substituted by a surrogate ingredient to replicate one or other of the solution's physico-chemical properties. What if the resulting bacterial challenge test fails, ie test bacteria are recovered downstream of the test filter? Does this imply that the process is no longer valid and that the product, safely used over many years, now poses a major health risk? Perhaps it would be better instead to say that the process no longer meets the original validation criteria, but this does not automatically imply a health risk as the process will originally have been approved under the terms of its marketing authorisation. How can the real risk of failure to sterilise by filtration be assessed, when the bacterial challenge used in the validation test (ASTM, 1983) is many orders of magnitude greater than that likely to be encountered at any time in the actual process?

VALIDATION — STERILE, IN PRACTICE

Validation of a pharmaceutical process to comply with GMP requirements (FDA, 1987; PDA, 1998; EC, 2004) is normally expected to be a forward-looking activity (Levchuk, 1994; Motzkau and Okhio, 2005). For new processes, individual process steps are designed and validated concurrently, prior to validation and implementation of the entire manufacturing operation. Even during the development stage, thought should be given to issues that might arise as the process is scaled up from the research bench through pilot and clinical trial stages, to full-scale manufacture (sometimes called *industrialisation*). Issues for consideration should include:

- Flow rate capabilities of the filter membrane and eventual size of the filter assembly.
- Membrane interactions with the active ingredients and excipients of the product formulation. Are there chemical interactions that damage the membrane, or extract substances from the filter materials into the solution? Do any of the solution's components bind to the surface of the filter materials?
- Proof of bacterial retention capability under simulated process conditions.
- Is a suitable filter device available at the intended final scale? For example, self-contained capsule filters may be convenient during process

development stages, but are limited in terms of their maximum available size and cannot be steam sterilised *in situ*. For the final process, is it intended to steam sterilise the entire system in place or would aseptic connections for one or more components be acceptable? Can a system with multiple capsule filters be assembled in a manifold and used safely without significant risk of contamination?

- Does toxicity of the product solution pose a risk to process operators or require special provisions for containment, making fully disposable process components desirable? Does this place a limit on the eventual maximum scale of the process? Do other physico-chemical properties of the product, such as pH, volatility, flammability, etc, constrain the choice of filtration equipment?

- Can the product solution be completely washed out of the filter with water enabling the performance of integrity tests after use under the correct conditions? Or can the filter supplier provide suitable parameters to test filter integrity in the presence of the product solution? Is the product solution of a suitable composition to enable the performance of such testing in a way that does not diminish the sensitivity of the test or compromise its validation?

- Can the filter supplier assist with validation studies during development and at full-scale implementation? Is the proposed validation programme up-to-date, so as to withstand current and forthcoming regulatory review?

Flow Rate

A filter manufacturer makes or buys membrane to incorporate into filtration devices. The membrane will have been made so as to comply with its manufacturer's defined performance specifications. These may include limits for allowable water flow rate capability, sometimes expressed as flow vs differential pressure, flow/pressure drop or simply flow/ΔP. When determining the correct size of filter assembly for an industrial process, this information is important to an understanding of on-going process filter performance. The filters used in validation of the process will represent some part of the possible range of filter specifications available from the chosen supplier, but where did they lie in the allowable range of values for flow capability? Were they from the more open (porous) range of allowable membrane batches or from the tighter (less porous) range? To what extent did the validation filters represent the entire range of filter specifications likely to be used in the process? Answers to these questions require open dialogue with the filter supplier, so that the width of the specification range in relation to the pharmaceutical process validation might be known.

The importance of these considerations becomes apparent when the implications of filter blockage are considered. A filter is blocked not, as might be insisted by a filter manufacturer, when the differential pressure exceeds a maximum permitted value of 4 bar, for example. A filter is no longer usable in a process when the flow rate through the filter has fallen below an acceptable limit value, such as that required to feed a filling machine, ensure completion of batch transfer within a specified time, or when the maximum available pressure from a pump or compressed gas has been reached. In the industrial process, how blocked is the filter at the end of the batch and how susceptible to change is this parameter as a function of the quality and *filterability* of product raw materials? Regulatory recommendations (FDA, 1987) required that filter validations and on-going use should include monitoring and recording of differential pressures across filters during and at the end of the process, and that significant deviations from the norm are investigated. However, this practice supposed that changes in differential pressure between filtration operations on different product batches risked compromising the validation of the sterilising capability of the process. Although this seems inconsistent with the general precept of a worst case approach to filter validation (PDA, 1998; FDA, 2004), it is nonetheless sensible that a significant deviation of a process operating parameter from the norm deserves investigation. In the context of filter blockage, the validation of the filtration process should therefore include consideration of the most and least *contaminated* product batches likely to be encountered, in conjunction with the most and least porous specifications of the chosen filter material. By this means it can be assured that limiting conditions of flow or pressure will not be incurred during process operation.

Membrane Interactions — Compatibility, Adsorption and Extractables

Microporous filtration membranes present a very high surface area in relation to their mass or volume, and therefore potentially render the membrane material more susceptible than bulk material to chemical interactions with the fluid being filtered. The very nature of filtration, where liquid flows through the membrane material, also ensures there is a continuously replenished source of potentially aggressive chemical at the filter surface. Consideration must be given to all components of the solution: active ingredients, excipients, trace materials (such as acid or alkali used for pH adjustment) and potential trace contaminants in these components. These latter substances are frequently overlooked, despite the fact that the filter membrane will retain and hence concentrate them at its surface.

The filter membrane material must be compatible with the solution undergoing filtration, in the sense that it must not be chemically attacked by the solution, leading either to addition of filter-derived substances into the solution or compromised bacterial retention capability. Compatibility can be assessed in a variety of ways. The most common approach is to perform a static soak of a filter element (or elements) in a sample of the process fluid at an appropriate temperature and for a period equivalent to the longest likely process duration. The filter is tested for integrity before this treatment, generally water-wet, and the test is repeated after the treatment, again water-wet. Two tests performed consecutively on the same filter will typically yield very similar results when the same conditions of wetting, equilibration and testing are observed. In this case however, some differences between these two integrity test results are likely, attributable to a number of causes.

- A small difference may represent the variation between the results of two tests performed on different days. It might also indicate a minor degree of chemical incompatibility, insufficient to compromise the functional integrity of the filter, but enough to affect the numerical result obtained in the test. This will become apparent in future tests for extractables and could have implications for the choice of filter material in the application.

- A larger difference might arise from the above and from variations in the degree of wetting of the filter membrane for the test. Typically a fully wetted membrane will give a lower diffusional flow test result and a higher bubble point test pressure than one that has some small region(s) where complete wetting of the porous matrix has not taken place. The same result may equally be obtained if the filter has not been completely flushed with water following the product exposure, and some traces of the product therefore remain in parts of the membrane. These effects can be overcome by using two wetting fluids and performing two tests of filter integrity — both before and after exposure to the product solution. One fluid should still be water (where appropriate for the membrane type) and the other an alcohol — propan-2-ol (or iso-propanol) is commonly used by filter suppliers. As it has low surface tension, the alcohol will eliminate difficulties of wetting that can be associated with water-wet tests, although caution should be exercised if the alcohol is immiscible with the product or causes precipitation of one or more of its components. If the observed difference in pre- and post-exposure integrity test results in water is not an artifact of the test, the same trend should also be seen in the results of the alcohol-based test.

- A very large difference, seen as failure of the integrity test, indicates either gross physical damage to the membrane or an inability to rinse the product solution from the membrane with water. The result should be confirmed using other wetting fluids, according to the filter manufacturer's recommendations.

Using a non-destructive integrity test method to assess filter compatibility has the advantage that it allows other tests to be done on the same filter element(s). For example, the filter(s) may be subjected to a bacterial retention test (see later in this chapter for discussion of these tests) to confirm that the sterilising capability of the membrane has not been compromised. But using the integrity test in isolation as a means to determine membrane compatibility only gives part of the necessary information. The filter supplier will have validated the test by correlation with bacterial retention efficiency. and not as a measure of the membrane's chemical and structural status.

Chemical compatibility of a filter membrane may also be evaluated using tests of physical strength, such as burst pressure or tensile strength on samples of defined size. Spectroscopic analysis of the membrane may also provide information about the chemical composition, since identification of new signals in HPLC, MS or infra-red analysis could indicate the presence of changed chemical entities following reaction with the product solution to which the membrane had been exposed. Other indices of membrane condition, such as melt flow index and determination of mean molecular weight may also be applied to polymeric membranes.

Two other test methodologies can be used to infer information about filter compatibility with a product solution, in the context of potential effects of the solution upon the efficiency and composition of the membrane materials. They are the determination of extractable substances from the filter, and demonstration of bacterial retention efficiency. The former is more sensitive in terms of chemistry, providing information about otherwise potentially-undetectable effects on the filter that would certainly result in product contamination. However, the latter — bacterial retention efficiency — is frequently seen as a more important characteristic, even in this context, because it represents the primary function of the filter. Both test methods are discussed later in this chapter.

As well as not contributing extraneous substances to the product solution, the filter should not remove components from the solution, with the notable exceptions of contaminating particles and bacteria, of course. The high ratio of internal surface area to volume of polymeric microporous membranes makes them susceptible to adsorption of molecular species from solutions passing through them. At risk of adsorption and of particular concern are the active ingredients and substances present in low concentrations, such as preservatives. The adsorption of even relatively small amounts of these substances can have a significant effect on their concentration in the product solution.

Adsorptive effects can be investigated during process development in laboratory scale experiments (Detyna, 1995) with the filter supplier's involvement in providing samples of appropriate filter membrane material and assistance with small-scale simulations of a process filtration. Logically, the experiments are generally best performed on the pharmaceutical manufacturer's premises, where there is access to samples of freshly made product solution. In addition, assays of the pertinent components of the product solution are best undertaken by the analytical laboratory services of the pharmaceutical manufacturer. The assay methods for many active pharmaceutical compounds may be quite complex, and the facilities there will have already been established and validated. The evaluations are best performed in the actual product formulation, and assays of the active ingredients may be complicated by the presence of other product components. In most cases for both parties the additional pain associated with validation of the necessary analytical methods by the filter supplier is not warranted; for a drug in development, both its chemistry and analytical methods may be highly confidential.

Different membrane materials have affinities for different molecular species, so it is not possible to extrapolate, for example, from one type of preservative to another. Two commonly used preservatives, benzalkonium chloride (BKC) and chlorhexidine (di-) gluconate (CG), are used in concentrations as low as 0.001–0.002%. One membrane type may adsorb significant quantities of BKC and have no affinity for CG, while for a second the converse may be true. Highly-active therapeutic biopharmaceutical molecules often have a strong tendency for surface adsorption and may be particularly susceptible to significant falls in concentration by these mechanisms.

Adsorption is an equilibrium phenomenon, in which molecules are attracted to and retained at the surface of the adsorbing material. The quantities adsorbed and the rate of adsorption are affected by factors including the concentration of the adsorbed molecule(s), the temperature, the presence of other molecular species and the affinity of the material for the target molecule(s). Adsorption is most noticeable at the start of the filtration, when all potential binding sites on the membrane are vacant. As filtration continues, the binding sites are progressively filled until all are occupied — the saturation point of the membrane. Thereafter a dynamic exchange can take place, in which each molecule released is immediately replaced from the solution. This is undetectable in practice, unless specifically evaluated using a solution containing labelled target molecules.

The presence of molecular species other than the target is important when evaluating the significance of adsorption for a particular product. Small-scale evaluations conducted using a solution of appropriate concentration in a suitable model fluid (such as water) may be convenient, but the kinetics of adsorption can differ from the same evaluation performed in the actual product solution, when other molecular species are present.

The impact of adsorption on a pharmaceutical process depends on the way in which the filtration operation is carried out. In processes where the product solution is transferred in bulk through the sterilising grade filter, from a preparation tank to a sterile holding tank, the effect of the initial adsorption of a solution component is rapidly obscured by the remaining volume of solution, so the adsorptive losses may be analytically undetectable. These losses may be compensated by the addition of an excess quantity of the adsorbed solution component, so that following transfer of the bulk solution the component is present at the correct concentration. Alternatively, the filter can be pre-saturated using an aliquot of the bulk solution that is discarded before filling the sterile tank. When sterile filtration takes place in the feed line to a filling machine, the first product containers filled will have a reduced concentration of the adsorbed component(s) and must be discarded. In some cases it may be possible to pre-saturate the filter by recirculating the bulk solution prior to commencing the filling operation.

Strategies for overcoming the effects of adsorption can also be evaluated with the aid of the filter supplier. Bench-scale tests will allow selection of the optimum membrane and a suitable strategy to prevent an impact on quality by adsorption of a key component of the product.

Filter Extractables

Substances extracted from a filter assembly by the product solution constitute a contaminant in the product. Regulatory and pharmacopoeial limits exist for some contaminants, such as particles and bacteria. Permitted maximum values for batch to batch carry-over could be considered in this light with respect to a maximum permitted level of unidentified substances in a finished dosage form. Beyond this, no limits are suggested for a permitted level of extracted materials from filter assemblies (or other polymeric components of a system, such as gaskets). The requirements of GMP and expectations of regulators are generally interpreted to mean 'as low as possible', with all reasonable steps taken to minimise the quantity.

Filter manufacturers and users took early comfort in the use of permitted materials listed and approved by FDA as suitable for food contact and packaging (FDA, 1993). Subsequently greater emphasis was put on the results of pharmacopoeial biological safety testing; although not intended for qualification of filters and other process components, these *in vivo* tests are relevant to pharmaceutical materials and medical devices (USP, 2005).

Filter suppliers have made data available in their own documentation and in published papers. Pharmaceutical manufacturers can use these data to estimate the impact of the filter assemblies used in their processes on the chemical composition of the end products.

Current practice for validation of filter extractables amounts to determination of the likely worst case mass of non-volatile extractable substances from the chosen filter, combined with a suitable pre-use flushing protocol to reduce the level of extractables to below detectable levels. The residue extracted from the filter is typically analysed spectroscopically, making a comparison with the signals obtained from untreated filter materials. Thus it is possible to say there are no new spectroscopic signals apparent, and therefore there has been no chemical interaction leading to the formation of new substances.

The analysis of extractable substances from filters is complicated by the need to understand what might be extracted in pharmaceutical product solutions — which generally consist of a solvent, most often water — with various dissolved solutes or other liquid components. After a period during which the filter is exposed to the product solution at an appropriate temperature, an ideal analysis would probably involve some type of spectroscopic analysis of the filter extract and concentration of the solution by evaporation to dryness, when the mass of residue can be determined and further spectroscopic analysis be performed.

The reality is often different, either because the product solution is not sufficiently volatile or because the solution contains interfering solutes. A solution of 5% (*w/v*) glucose in 0.9% (*w/v*) sodium chloride (5% dextrose saline) contains 50 grams per litre of glucose and 9g/l of sodium chloride. Evaporation to dryness of one litre of the solution will therefore yield a residue containing 59g of solute. By comparison with this mass of 59,000mg of residue, the typical extract of between 10 and 50mg from a 25cm filter cartridge is of little significance and analytically undetectable. Whether analysed by spectroscopy in the raw extract or in the concentrated residue, the background signals from solution components will generally be strong enough to mask the signals from any filter-derived substances.

An answer to this question of filter extractables is to use a solvent model to imitate the product solution, in which all the components are volatile liquids that can be evaporated without leaving a significant residue. Background noise is further reduced by the use of especially pure analytical or spectroscopic grade solvents. The model solvent approach (Stone et al, 1995; Reif et al, 1996; Weitzmann, 1997) requires an examination of the chemistry of the pharmaceutical solution components (both active ingredients and excipients), to enable appropriate choices of volatile substitutes to be made. Consideration should be given to the overall nature of the molecules and to the chemistry of specific functional groups present on each, so that a suitable solvent model can be designed. Classifications of solvents according to their solvating power (Rohrschneider, 1973; Synder, 1974; 1978) have been used to support this strategy. Some approaches to the design of solvent models have proposed

(Weitzmann, 1997) that solutes present at less than 5% or 10% concentration in the final product should be disregarded, on the grounds they do not constitute a significant presence in the solution. However such an approach may ignore the potential effects of strong acid or alkali groups, alcohols and other solvent types that may extract substances from the materials of filter construction.

The choice of solvent model must be justified by reference to the chemistry of the product solution. The rationale for choice of each component of the model is based on the chemistry of functional groups common to the model and the product solution. It is one thing to argue that a 4% concentration of an alcohol functional group is negligible in the context of the quantity and identity of filter extractables; it is another to include this functional group in the solvent model and prove it.

The analytical methods employed must be chosen with care. Significant emphasis is given to chromatographic methods (PDA, 1998) such as gas chromatography and high performance liquid chromatography (HPLC), but these methods are comparative, relying on choice in advance of appropriate standard molecules and solutions. This author has encountered instances where substances not used in or associated with the manufacture of a particular filter cartridge are claimed to have been identified in the residue extracted from these filters. Further examination, for example by selecting a different chromatography medium or solvent system, confirmed the coincidence of a potential filter extractable material that co-chromatographed with the selected standard solution under one set of analytical conditions.

The end result is that the pharmaceutical manufacturer knows that some material is extracted from the filter and how much of it there is (quantity), knows where it came from (identity), and can provide assurance that the material meets the requirements of tests for biological safety (toxicity). Finally, the pharmaceutical manufacturer has taken suitable steps to minimise the presence of this material in its final product, by means of suitable washing procedures.

Bacterial Retention Testing

Sterilising grade filters are qualified by their manufacturers using variations of a standard test method (ASTM, 1983) published more than 20 years ago. This requires complete removal of a minimum challenge level of 10^7 colony forming units (cfu) of the test bacterium *Brevundimonas diminuta* ATCC19146 (formerly *Pseudomonas diminuta*) per square centimetre of membrane surface area. This bacterial level was proposed as a means to ensure that a sufficient challenge was given to the membrane so that every *pore* is challenged and given the opportunity to allow passage of the test organism. However, whether there really are 10^7 pores per cm^2 of filter area is largely irrelevant to the use of this challenge level, since significantly higher challenge levels are impractical (they might block the membrane and prevent completion of the test). In any case, a challenge level of 10^7 bacteria per cm^2 of effective filtration area is orders of magnitude higher than would be encountered in a sterilising filtration process. The American Society for Testing and Materials (ASTM) method (ASTM, 1983) justifies this challenge level as providing 'a high degree of assurance that the filter would quantitatively (*sic*) retain large numbers of organisms.' The importance of this concept is emphasised by reference to the 'requirement to provide a quantitative assessment in validating a sterilisation process', and is followed by a description of how a numerical value for filtration efficiency might be used to determine the probability of obtaining sterile filtrate or indeed meet a specified probability of sterility assurance. Regulatory compliance guidelines subsequently espoused (FDA, 1987; 2004) the idea of the proposed minimum challenge level of 10^7 bacteria per cm^2 of filtration area, which is now widely accepted (PDA, 1998). It is generally agreed, however, that a numerical value such as an SAL, can never, under any circumstances, be assigned to the end products of a sterilising filtration process. (An SAL is calculated for distinct product units — bottle, bags, vials — with previously-known maximum bioburden levels, and equally cannot therefore be applied equally to a system sterilised by a steam in place procedure.)

A manufacturer's qualification data for sterilising grade filters are typically published in the form of a supplier's Validation Guide. The pre-eminent purpose of these documents is to present data demonstrating the ability of the filters in question to retain large quantities of test bacteria under laboratory conditions and to correlate this capability with the results of non-destructive integrity tests of the filters. The approach taken to integrity testing and the choice of supplementary data in the documents are matters of individual preference on the part of filter suppliers. What is certain is that these data, once considered sufficient in themselves, have for many years no longer been adequate to validate a sterilising filtration process for a pharmaceutical product. Indeed, as the market for sterilising filters has matured, so the documents produced by different suppliers have come to resemble one another, and they now scarcely even serve to distinguish between suppliers as an aid to filter selection.

Currently the most important aspect of validating bacterial retention efficiency concerns process-specific validation. The chosen filter is tested using the procedures described above, but the operating conditions of the test are changed to simulate the actual pharmaceutical manufacturing process for a particular drug product. The purpose of this is to take into account all possible factors and interactions that may exist between the filter components, the mode of filter action, the physical conditions of the process and the physico-chemical characteristics of the pharmaceutical product solution. This brief description is enough to illustrate the enormity of the proposal, in fact that a sterilising filter might only be considered genuinely sterilising grade for a particular product and process after it has been validated for that specific product and process. Furthermore, a change in the total volume of the batch or flow rate of filtration, for example, might render that validation obsolete and necessitate repeating the study using appropriately modified conditions. That indeed is how it is (FDA, 1987; 2004).

Some efforts were made to reduce the obvious scale of work involved (Levy et al, 1990) by pooling data from a number of studies in a matrix, so individual process or product parameters could be said to have been validated, avoiding the need to repeat testing. The approach received fleeting regulatory approbation in a policy guidance document (FDA, 1994(a)), and was even described, after intensive evaluation, as 'scientifically sound and acceptable', so long as a report showed 'the characteristics of the product in question and the products (*sic*), from the database, used to show that the filter will sterilise the product'. Very shortly afterwards this acceptability was rescinded (FDA, 1994(b)), not, after all, having been intensively evaluated. In the absence of supporting data it was recommended that microbial retentivity is tested by microbial challenge in the actual drug product. Nonetheless from these events it emerged that testing of all members of a family of similar or related products might not be required. For example, where a range of different concentrations of the same active ingredient were concerned it was possible to bracket products and test the extreme concentrations. In the same way, products with similar physical properties or process parameters might be grouped for the purposes of bacterial retention validation. An important distinction is made between bracketing of data and unscientific extrapolation (Levy et al, 1990; FDA, 1994(a), (b); Madsen, 1995). This has important implications for retrospective validation of existing processes, although only reducing the burden of work in one aspect.

The term *parametric approach* has been used (Martin and Brantley, 1994; Docksey et al, 1999) to describe the alternative and self-evident approach to process-specific validation of sterilising grade filters — that is to consider each combination of product and process as unique unless there are compelling reasons to justify a grouping or bracketing approach for a product family. For new products in the development stages or undergoing transfer to industrial scale manufacture, this is the only feasible approach, since usually there are no other products with which it might be jointly considered.

Process Simulation — Broth Media Fills

With recent changes to guidelines on procedures that demonstrate good process control (FDA, 2004; EU, 2004) it is useful to consider the art of simulations for aseptic filling processes, sometimes referred to as *broth media fills*. Some aspects of the current requirements have been well covered, but others are less clearly understood (Booth, 2006; Halls, 1995).

Media fills simulate all events surrounding an aseptic filling operation for sterile liquid products. This process normally includes a sterilising filtration step, either to sterilise the product or to provide final assurance of sterility before filling into the final container. By (filtering and) filling liquid culture medium, the risk or probability of contamination of the filled product can be determined. It was therefore often considered appropriate to prepare the required volume of liquid culture medium (such as tryptone soy broth) and pass it through the normal sterilising filter and into the filling process. Furthermore it has sometimes been the case that this methodology was believed simultaneously to validate the sterilising filtration step. Not so, for reasons of approach discussed above.

This naive concept of a media fill has some serious flaws. First, the dehydrated culture media are normally supplied non-sterile and carry a bioburden at or above 10^4 cfu/g, equivalent to about 2×10^4 cfu/mL in the reconstituted broth medium. This renders broth media more highly-contaminated than is permitted for aseptically filled products prior to filtration (not more than 10cfu/100mL (EMEA, 1997)), and therefore evidently unsuitable to be taken into a clean manufacturing area adjacent to an aseptic filling suite. Additionally, the bioburden in this broth probably includes a wide range of bacteria not previously encountered in the manufacturing equipment, thereby introducing new contamination to the facility. Broth media typically require heating to ensure they are fully dissolved, a practice that takes time and energy, and that is not frequently carried out on a large scale. Not dissolving correctly, the broth medium was not readily filterable, and would result in blocking the process filters during the media fill and thus failure of the exercise.

One means of preventing these problems has been to use pre-sterilised broth medium for the process simulation. The broth would be sterilised by heating in a tank to 'autoclave' it, or by filtration in an entirely separate operation. The vessel of sterile broth medium could then be brought to the filling suite and connected aseptically to the filling machine at the start of the process simulation. However the liquid product filled in an aseptic process would normally be formulated or transferred into a non-sterile tank before filter sterilisation into the filling area. Such a procedure of broth pre-treatment might be interpreted as not fulfilling the spirit of the revised guidelines.

A more thoughtfully-designed media fill employs cold-filterable dehydrated broth media that have been pre-sterilised by gamma irradiation. The broth powder is reconstituted with purified process water or *Water for Injection(s)* (WFI), in the same way as the actual product might be formulated. Thus the broth medium has a similar chance of contamination as the product, and is not at risk of blocking the sterilising grade filter(s) in the process. The broth medium goes through the same filtration step as for the product and, while this does not validate the filtration, it does ensure that the process risks inherent to the filtration step are included in the process simulation (Booth, 2006).

Filter Device Suitability

Capsule filter devices are convenient to use, especially on the laboratory bench or in small-scale processes. They are compact, can be easily autoclave sterilised and do not require separate steel filter housings to contain them. Development of improved filtration materials and construction techniques has made capsules more suitable for a wider range of processes than ever before, as long as a sufficiently large filter device is available to provide the process capacity required at full scale. But filter capsules cannot be steam sterilised in line — they must be autoclave sterilised and subsequently assembled aseptically into a process system. While in many cases this may be achieved with relatively little risk, the pharmaceutical manufacturer must assess the risks associated with these aseptic manipulations as a function of the particular steps in the process where filtration is employed. Where an adequately large filter capsule is unavailable, a manifold of two or more capsules might be considered as a means of increasing total throughput capacity or flow rate to meet the demands of downstream processes.

Hazards associated with the product solution, such as toxicity or chemical nature, may make management of steel filter housings more difficult, as they pose the risk of operator exposure to the solution. When the process batch size is small or the solution relatively clean, and where there is no requirement for a high flow rate, a small capsule filter may be adequate. The enclosed nature of the assembly is ideal, as it can be handled more easily after use without creating a risk of product exposure to process personnel.

Sterilising grade filters must be tested for integrity after use (FDA, 1987; PDA, 1998; EC, 2004; FDA, 2004). Currently-available test methods for liquid filters require the filter membrane to be thoroughly wetted. Filter manufacturers validate their particular integrity test methods by correlation between the results of their test methods and the results of bacterial retention tests, using water as the membrane wetting fluid. It is not always appropriate to introduce water into a process system, either before or after filtration of the product. This may be because water is undesirable in the product or there can be no risk of product dilution, or it may be because the product cannot easily be flushed out of the filter after use. In these instances a filter user would prefer to test filter integrity using the product solution itself as the wetting fluid. Such test parameters are available from all major filter suppliers, and are determined according to their own interpretation of available approaches (PDA, 1998). This is also dependent on whether the composition of the product solution makes it suitable for the determination of integrity test limits; highly volatile solution components may mean that the appropriate test limits do not provide for an adequate level of test sensitivity.

INTEGRITY TESTING — STERILE, IN PRINCIPLE

Methodology

The methods employed for testing the integrity of sterilising grade liquid filters have remained broadly unchanged in concept for more than 25 years and are well described elsewhere (PDA, 1998). In summary, a liquid-wet membrane filter is put under a pressure of compressed test gas, usually air or nitrogen, and the resulting gas flow determined. One methodology relies on retention of the liquid by capillary effects in the pores of the microporous membrane, so that a certain pressure must be exceeded before the liquid is expelled. This 'bubble point' pressure is related to the diameter of the pores in the membrane, as well as the surface tension of the liquid and factors relating to the uniformity and configuration of the pores. The higher the pressure required to displace liquid from the membrane, the smaller in principle the size of the pores. The alternative methodology is based on the solubility and diffusion of the test gas in the wetting liquid. At a constant applied pressure, lower than the 'bubble point', test gas under pressure on the upstream (inlet) side of the wet filter membrane dissolves in the wetting fluid and diffuses through the film of wetting liquid in the filter's pores. The gradient of elevated gas concentration subsequently reaches the downstream (outlet) face of the membrane, where the ambient gas pressure is equivalent to atmospheric. Here the gas comes out of solution and the volume of gas flow can be measured as a function of time.

Each method has its promoters and detractors. The 'bubble point' of a membrane is independent of membrane thickness, and so may be considered more suitable for testing the integrity of a filter designed to remove particles and bacteria by size exclusion mechanisms alone. It is, however, more widely accepted that sterilising grade filters rely on both their 'porosity' and membrane thickness to achieve their desired function (Osumi et al, 1991; Pall and Kirnbauer, 1978).

The diffusion-based test method employs a test gas pressure approaching (but lower than) the 'bubble point' pressure, to allow identification of filters with oversize or defective pores. The rate of diffusion (described by Fick's Law) is proportional to the distance over which it takes place, in this case the thickness of the liquid film wetting the membrane. In both test methods, the results can be measured from either the upstream (inlet) or the downstream (outlet) sides.

For gas sterilising filters the same test methodologies can be applied. One important difference is that these membranes are hydrophobic in nature and cannot be wet with water or aqueous solutions for the purpose of these tests. Manufacturers of these filters recommend the use of various alcohol-based solutions in this case. Some suppliers recommend aerosol-based tests for gas filters; however these methods have generally found greater acceptance for coarser gas filter screens and HEPA-type filters panels used in laminar flow and air-conditioning systems.

A significant change in filter testing methodology has been the development of a water-based intrusion test for hydrophobic membrane filters. This relies on the converse of the diffusion-based test principle for hydrophilic filters, that these membranes repel water and do not allow it to enter the membrane pores. There was initial debate over the actual mechanism by which the observed test phenomenon took place, but it now seems reasonable to accept that the test gas (again, air or nitrogen) diffuses through the liquid film under the influence of the applied test pressure. This time, however, the gas comes out of solution at the upstream (inlet) face of the membrane and then passes through the filter pores to reach the downstream side where it escapes. In practice, the measurement of results can only be performed from the upstream (inlet) side, because the very small gas flows that take place require a bubble or film flow meter, a reference device not suited to routine measurements.

Practice

Great improvements in measurement methods and operation of these tests have taken place over the last 25 years. From manually-operated test kits using a pressure gauge and regulator, automated machines using pressure transducers have been introduced and brought with them a number of benefits. The most notable improvement was the elimination of operator dependence, since the machine performed tests in the same way every time, so a consistent procedure and interpretation were applied. Pressure transducers improved the accuracy and sensitivity of the measurements made, improving reproducibility of the test.

The measured phenomena, either diffusion or breakthrough of air to the downstream side of the filter under test, were initially appraised on the downstream side. This posed problems under aseptic process conditions when the filter and downstream system were sterile and unnecessary interventions were best avoided. It was possible to perform diffusion-based tests from the upstream side of the filter using manually operated test equipment, thus avoiding intervention in the sterile side of the filter. A mathematical formula is used to transform the required downstream diffusion test limit into a corresponding decay in pressure on the upstream side (PDA, 1998). The calculation also requires knowledge of the volume of compressed gas isolated on the upstream side of the filter, and the test itself requires essentially constant temperature throughout the measurement phase. The practicality of these tests was limited by the sensitivity of the pressure gauge used. Typically a 15cm diameter gauge with full-scale reading of 6 barg has graduations of about 2mm representing 50mbar pressure increments. At best the test was therefore likely to distinguish between a filter that was an evident failure and one that was probably satisfactory. The automated machines use pressure transducers that, when calibrated, can be relied on to provide reproducible sensitivity to 1mbar over a range of 6–10 bar pressure. It is this dramatic improvement in test sensitivity that enabled development of the water intrusion test for hydrophobic filters, since the small changes in pressure observed during this test could not have been meaningfully interpreted using manual test equipment. The pressure sensors in these machines also enable measurement of the upstream volume of the test filter system as part of the test cycle. The machine can then be programmed with the supplier-validated diffusion test limit; the software will perform the required calculation to convert the measured test result back into a diffusion value for comparison with the supplier's limit value.

Further developments have taken place in measurement technology for filter testing, using two methods of flow measurement from the upstream side of the filter, avoiding the need to know or measurement of the upstream volume of the test system. One is the application of sensitive mass flow meters. These instruments are generally best used as reference devices and are less well suited to routine measurements in a processing environment. The ability of these filter test machines to make real-time measurements of flow throughout the test period provide more data than had been previously available from filter tests, so stability of the test result could be inferred. An automatic test facility was also offered to enable the machine to determine if the test result was statistically stable and so save time by discontinuing the test before the end of the pre-selected test duration. The potential for interpreting the additional data available and the greater susceptibility of these machines to environmental fluctuations risked the re-introduction of operator dependence to filter testing. Supply of these machines was discontinued only a few years after they were first introduced. A parallel development was a flow dosing technology, in which the upstream side of the filter system is held at the correct pressure by small volumes of compressed gas dosed into the system. The amount of gas introduced to maintain upstream pressure is calculated using the Gas Laws by reference to the known volumes of small pressurised reservoirs inside the machine. Once again, the technology gives a real-time view of the progress of the filter test.

Both technologies require more sophisticated software algorithms to interpret the greatly increased amount of available data and users should have a clear understanding of the manipulations involved to satisfy themselves of the validity of the results obtained.

Increasing automation of all aspects of aseptic processing has lead to greater use of electronic data storage and many pharmaceutical manufacturers retain increasing amounts of product batch data in electronic form. The same approaches have been applied to these data forms as to paper records — the need for clear audit trails, signatures and immutability of the finished record. Some filter test machines comply with the requirements of the 21CFR part 11 guidelines on electronic signatures, while others do not. However it might be more appropriate to say that some filter manufacturers claim compliance while others do not. The difference in interpretation of this point is that none of the filter test machines currently available are intended to act as long-term storage devices for the final test records produced. All offer facilities for transfer of the test results into the host data storage systems of pharmaceutical manufacturers for long-term storage. It may therefore be the case that it is here that the requirements of 21CFR part 11 regarding electronic signatures and data security should be applied, and the relevance of these guidelines to filter test machines becomes open to debate.

The pressure decay measurement of filter integrity requires conversion of a validated downstream flow limit value to an upstream pressure decay limit value. Flow values in the downstream test are measured at a constant input pressure of test gas and are validated in this way by suppliers. The pressure decay test is initiated at the same pre-determined test pressure, but pressure necessarily falls during the measurement phase of the test. The rate of gas diffusion through the filter is proportional to the applied gas pressure (Fick's Law, (PDA, 1998)), therefore the decay in pressure must necessarily follow an exponential decay algorithm. Some filter suppliers provide the necessary test limits based on the assumption of a linear pressure decay algorithm, yielding potentially unsafe test limits — larger values of allowable pressure decay than are actually representative of the corresponding diffusion flow limit under the designated test conditions. Pharmaceutical manufacturers should satisfy themselves as to the suitability of the limit values provided and their means of determination.

Testing Regimes

One purpose of a filter integrity test is to ensure that, once a filtration process is properly validated, identical filters (pore size rating) are used in production runs (FDA, 2004). The test confirms consistency in the performance specification of the filter with its original validation as sterilising grade, and with the process-specific bacterial retention validation studies (FDA, 2004).

Current GMP (EU, 2004; FDA, 2004) requires a test of filter integrity to be carried out after use of the filter to assure sterility of the pharmaceutical product. There is no debate as to the need for this test. The result of the test forms part of the documentation used to justify the release of the product batch from production into the market. An integrity test confirms that the filter has not been damaged by handling during installation, product exposure, process conditions or excursions, and other treatments since the last such test was performed. The result is particularly important where, following filtration, the product has been filled into the final container. In the case that the product has been filtered during transfer into a sterile holding vessel, an out-of-specification result for the filter integrity test may trigger a validated procedure for batch recovery, such as a second transfer taking place within a specified period through a second sterilising filter and into another pre-sterilised vessel. For product already filled into final containers, this may be re-filtered and re-packaged — if economically viable — and if final product quality is not compromised by the necessary procedures.

Some regulatory guidance proposes that integrity testing of sterilising filters *can* be carried out before processing (FDA, 2004), while other guidelines (EU, 2004) indicate the desirability of a filter test after sterilisation but before use. This pre-use test gives both process and economic benefits. It confirms that the filter is within specification, of the correct grade and undamaged by prior handling and sterilisation treatments, so that product can safely be committed to the process. It also obviates any concern that a filter damaged prior to use might become partially plugged during use so as to block the site of damage and obscure it from detection in the integrity test performed at the end of the process. This is not to say that damage incurred *during* the filtration may not subsequently be *repaired* by blocking of the membrane in this region. In the event that a filter has been damaged before processing of the product batch, testing filter integrity prior to processing also avoids the waste of time and resources associated with such a *false start*.

A pre-use filter integrity test also aids good management and manufacturing practice, since the identification of damaged filters prior to use enables investigation and identification of procedures leading to the damage, and hence elimination of some risk from the process. Up to 80% of all *genuine* integrity test failures of sterilising grade filters can be attributed to damage caused during steam sterilisation of the filter assembly (Martin, 2006), predominantly when steaming filter assemblies *in situ* on the process line. If filters are regularly damaged by a steam sterilisation procedure, then the latter can be said to be out of control and a source of risk to the process. In addition to the costs incurred in wasted filters and production time spent preparing and re-preparing the line, the risks are to the product itself, which may have limited stability prior to sterilisation by filtration: if too long a delay is incurred, the product batch may have to be scrapped.

Validation of Filter Test Limits

A sterilising grade filter which meets the manufacturer's specification for an integrity test is expected to provide sterile effluent and, once validated for the specific product and process, the test result is taken to guarantee the sterility of the product (all other aseptic process controls are here assumed to be satisfactory). Many filter integrity tests are carried out in the process environment, which is quite different from the conditions under which the filter manufacturer validated its integrity test. Furthermore, many pharmaceutical manufacturers use filter test parameters specifically provided by the filter supplier for use with their own product solution as the wetting fluid. Methods by which such values may be determined have been summarised elsewhere (PDA, 1998).

The use of filter test parameters that apply to the filter under test when wet with the pharmaceutical product solution has many benefits. There is no need to introduce another fluid into the process system before filtration, which risks dilution of the product. There is no need to rinse the filter with water after the filtration in order to remove all traces of product solution from the filter (nor indeed to validate the flushing procedure needed to achieve this). It is also possible to perform integrity tests during the filtration process, for example as part of the definition of sub-batches or during very extended filling operations such as those on Blow-Fill-Seal machines lasting 10 days or more, and filling several thousand litres in volumes as small as 2ml.

What are the important considerations for these integrity test parameters? Like all aspects of aseptic processing, there should be an associated safety margin between the users test limit and the point of likely failure to sterilise. The magnitude of the margin differs between manufacturers and is affected by many factors, including:

- The characteristics of the membrane.
- The conditions and type of test.

- The nature of the wetting solution.
- The size of the filter assembly or number of filter elements.
- The stability of test conditions (pressure, temperature) during the test.
- The relationship between the test conditions for the actual filter test in the process, and the conditions under which the original validation of the integrity test method was executed.
- The accuracy, calibration and validation of the automated device used to perform the integrity test.
- The methodology and conditions used by the filter supplier to determine product-specific integrity test parameters, and how they are linked to the originally-validated test limits for water or other standard wetting fluid, as published by the manufacturer.
- The degree to which the samples of pharmaceutical product solution provided for evaluation were representative of the current solution made and filtered at industrial scale.
- The permitted limits of variation in the test conditions during any particular test of filter integrity and between one test and another. For example, temperature, product concentration.

In consequence of the small and indefinable differences between one filter cartridge and another, differences will arise in the numerical value of results between one test and another. In most cases, it is sufficient if the test result is within the specification provided by the supplier to record a satisfactory result. However, to do so requires that the factors above have been taken into account and the test conditions judged to be valid.

In some applications the same filter is used repeatedly, such as the sterile filtration of large volume parenteral solutions prior to filling and autoclave sterilisation or filling of some ophthalmic solutions. All such uses must be validated to the limit of the number of repeat uses envisaged. The most frequently reused filters are gas filters used as the sterile barrier on the vent of product holding tanks and sterile gas supplies used to provide top pressure to vessels. The frequency with which such filters are integrity tested is a matter for the individual user. Where a filter is tested on a batch by batch basis the user can be sure in the release of each batch of product. If a filter is tested only after every tenth batch, for example, the manufacturer might hold those 10 batches in stock until a satisfactory test result is recorded, or might release the product and risk making a decision about product recall in the event of a filter failure. The gas filters are not considered as being in direct product contact, although the gases they filter may be.

The test results for filters that are repeatedly used and tested should be evaluated for trends. An upward trend in diffusional flow or pressure decay value, towards the test limit, may indicate a gradual deterioration in the filter membrane mediated by the repeated steam sterilisation treatments undergone. A sudden increase in the test result, although still within the allowable specification, may be a sign of incipient filter failure; the filter should be discarded. Progressive falls in the numerical value of the test result could indicate that the filter is starting to block. These trends become important as data are collected, as they provide a means to monitor the process over time. This may provide information about the process but should also be shared with the filter supplier, as it may be indicative of changes or trends in the filters themselves.

PENETRATION OR GROWTHROUGH — FACT OR FICTION?

A filter previously qualified as sterilising grade may be put in service for an extended period, such as in a water system or long-term product filling application. Here it is used for a week or more between sterilisation or sanitisation procedures and subsequent contamination of the downstream system may be detected. The contamination may have originated at a point of use and spread back upstream against the normal flow direction, suggesting poor management of environmental conditions at the point of use. On some occasions, however, the contamination is presumed to have arisen as a result of bacterial passage through the sterilising filter, a claim that diverts attention towards a system component that is less well characterised and whose performance is regarded as the responsibility of an external supplier.

The organisms isolated in such instances are typically small bacteria (Blosse and Sundaram, 1999; Howard and Duberstein, 1980; Sundaram, 1997; Ball, 1999), commonly of the genus *Pseudomonas* or closely related to it, such as *Ralstonia sp* (eg *R. picketti*) or *Hydrogenophaga pseudoflava*. Such filter *penetration* events involving small contaminating bacteria have become known as *growthrough*. This term derives from the supposed mechanism of penetration: bacteria multiply by a process of binary fission, a cell increases in size and then splits into two discrete entities. However, to be plausible it must be supposed that a single bacterium

Figure 6.2 Could this be bacterial 'growthrough'?

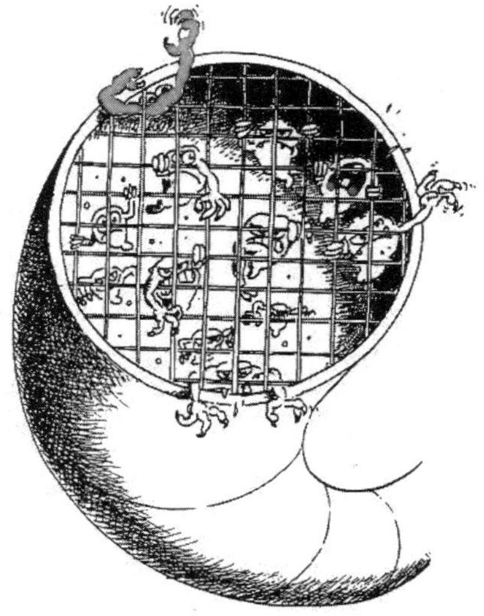

Penetration: the time-related emergence of bacteria downstream of a sterilising grade membrane filter, exceeding the filter's retention capability with a high challenge level of small bacteria ...

... or is it *Growthrough*?

retained by the filter membrane, possibly at or near the upstream surface, divides repeatedly until its progeny have spanned the thickness of the membrane. A bacterium in this low-nutrient environment may be 1.5µm in length while a typical sterilising grade membrane is about 175µm thick. The concept of growthrough therefore suggests something in the range of 100 or more consecutive bacterial divisions in the same plane through a linear capillary pore (which does not exist) in the membrane — an evidently ludicrous concept of longer duration and more complex mechanism than a simple ability to pass right through the membrane. In theory a bacterium with a division time of 30 minutes might achieve the required number of divisions in a little over 48 hours, but in practice insufficient nutrients are available to support such a rapid and sustained rate of growth.

If the original bacterium was attached or retained within the filter matrix, it is possible to speculate that the unrestrained progeny of a division might breakaway from its parent and travel further into the depth (or thickness) of the filter membrane. The *parent* bacterium would already be in close proximity with the filter material or located in a blind pore (or one small enough to have retained it at the outset), nonetheless it is feasible to suppose that such an event could occur and might be referred to as *wriggle-through*. The stream of fluid flowing through the filter would carry the *newborn* bacterium towards its logical objective of reaching the other side of the membrane, until the bacterium was retained in the filter matrix by direct interception or inertial impaction. Once again, for this to happen the bacteria must be in a growth phase, implying the presence of nutrients.

According to commonly-accepted definitions (FDA, 1987, 2004; PDA, 1998), the principle underlying the qualification and evaluation of a sterilising grade filter is that it should be capable of completely retaining the challenge of a conditioned diminutive bacterium (*Brevundimonas diminuta* ATCC19146) at a level at or above 10^7 bacteria per square centimetre (10^7 cfu/cm^2) of filter membrane area. A filter that allows passage of bacteria in this test cannot be classed as *sterilising grade*. These tests are normally carried out under laboratory conditions and within reasonable time constraints (such as the duration of a working day) and the recorded passage of bacteria is labelled *penetration*. A filter's ability to completely retain a challenge of a larger bacterium such as *Serratia marcescens*, even at a level of 10^7 cfu/cm^2, is not held as an indicator of its ability to retain the smaller *defining* organism *B. diminuta*, and will not qualify the filter as sterilising grade. A filter so qualified will reproducibly allow passage of *B. diminuta* in significant numbers in a laboratory challenge test, an observation satisfactorily interpreted as penetration. Yet curiously, when an analogous event occurs involving smaller organisms present in lower concentrations, it has been found necessary to invoke this mysterious concept of *growthrough* to account for it.

Circumstances surrounding these reported growthrough events can be explained by a mechanism of *penetration*, that is, the ability of a small organism to pass through the filter; there is no need to invoke special mechanisms. In a laboratory challenge test of a filter, penetration occurs because the statistical ability of the filter to retain the challenge organism is exceeded by the number or level of organisms challenging the filter (Pall and Kirnbauer, 1978). This principle underlies the recommendation to use 0.45μm membranes for a sterility test (USP, 2002), in which the analysis membrane filter is regarded as having adequate removal capability to retain at least a significant proportion of the contaminating bacteria contained in test samples.

For a 0.45μm filter qualified using *Serratia marcescens*, a challenge of *B. diminuta* at a level approaching 10^5 cfu/cm^2 is generally sufficient to result in detection of the challenge organism downstream of the test filter. A similar result would be observed whether the challenge were applied over 30 minutes or 30 hours, as the applied challenge level builds and the capability of the membrane is overwhelmed. The same principle applies to passage of small bacteria through a '0.2μm' filter used for extended service in a water system. The bacteria are smaller than *B. diminuta*, perhaps as a result of the nutrient-poor environment, and they are present in low numbers. It quite reasonably takes a considerable volume of water flow (and therefore by definition considerable time) before the challenge builds up to a level sufficient to overwhelm the filter's ability to remove them.

Broth media simulations of aseptic filling processes are also at risk from penetration events when the procedure involves sterilising filtration of bulk untreated broth medium. The powdered culture media commonly-used, such as tryptone soy broth, may contain a sufficient bacterial bio-burden to yield up to 10^5 cfu/ml in the re-hydrated broth. Broth media fills generally take place over several hours at most, rather than several days. However, at such high levels of bacterial bio-burden the resulting non-sterility of filled broth medium is hardly surprising. Once again, such a contamination event would be termed penetration, on the grounds that insufficient time had elapsed to suppose that the bacteria had either *grown* or *wriggled* through the membrane. Evidently in such a nutrient-rich environment the bacteria might be expected to reproduce rapidly, however media fill simulations are rarely of sufficient duration to permit the mechanisms outlined above.

The delay between the start of filtration and the onset of bacterial penetration through the membrane relates to accumulation of the challenge in the filter membrane, and not to bacterial replication spanning the membrane. A published study (Leo et al, 1997) employed small bacteria originally isolated from a long term filling operation, using Blow-Fill-Seal technology. The bacteria were cultured and pre-treated under conditions replicating the process environment from which they had been isolated. They were able to penetrate a 0.2μm sterilising grade filter (qualified using *B. diminuta*) when the challenge reached a threshold level and they were completely retained by a finer 0.1μm grade filter. Furthermore, when two 0.2μm filters were challenged in series, bacteria penetrating the first were equally able to penetrate the second, without the need for an extended period to permit growth, nor was there need to postulate exotic mechanisms of *growthrough* to explain the events.

The idea of growthrough was also examined from a different starting position, that it must occur and can be demonstrated (Cole and Parker, 1985). Sterilising grade filter cartridges (25cm long) were autoclave sterilised in an inverted position (open-end uppermost) in glass cylinders. Each cylinder, containing a filter with a protective cover, was placed in a protective laminar flow cabinet; the spaces inside and surrounding the cartridge were filled with sterile culture medium (tryptone soya broth). The broth surrounding the filter cartridge was inoculated with a culture of *B. diminuta* and the cover was replaced. After a short incubation of about 48 hours the broth medium surrounding the filter cartridge was turbid with bacterial growth. During a period of one week, daily samples of broth were withdrawn aseptically from the inside of the filter cartridge and incubated; all remained sterile. At the end of these experiments, the bacterial titre surrounding the filter was typically in excess of 10^{10} per millilitre, yet the broth medium contained in the core remained sterile.

Despite ample opportunity of both time and nutrient source, growthrough did not take place. It might be concluded that either growthrough does not occur or that this examination did not create the right circumstances to enable it. All incidents of bacterial passage thus far encountered in the experience of this author could be explained on the basis of a time-dependent and hence challenge level-dependent penetration. Furthermore, while some events could be satisfactorily explained by a combination of *penetration* and/or *growthrough* and others only by *penetration*, none were capable of rational explanation on the basis of *growthrough* alone. The term growthrough is therefore erroneously used to explain any penetration event that shows a time dependency, however in filtration time relates directly to the volume filtered and longer filtrations correspond to higher challenge levels. So penetration of the filter may simply be the result of exceeding the filter's retention capability.

DISPOSABLE SYSTEMS — CAPSULE FILTERS AND MUCH MORE

Disposable Capsule Filters

Filter assemblies traditionally comprised stainless steel housings containing one or more filter cartridges. In the early days of the biotech industry small capsule filters (Martin, 2005) were used as vessel vents and for filtration of small volumes of buffer or product solutions. A filter capsule is a self-contained filter assembly made up of a filter element within an outer shell or casing, the capsule. Early designs were restricted in size and application because of limitations placed on their construction and use by the materials of construction. The capsules were generally made using polymers similar to the filter element contained within, thus avoiding the use of additional materials. However the desirable size of a capsule and the strength that could be achieved for its walls was compromised.

Capsule filters could be sterilised by autoclaving but early polymers were not sufficiently stable to allow sterilisation by gamma irradiation. Today a range of gamma-stabilised polymers, such as polypropylenes, polyesters and polysulfone materials, are available. These contain additives that enable them to withstand gamma irradiation without subsequent degradation.

Process and circumstances determine the reasons for using capsule filters. They can offer a variety of advantages in the right conditions, including elimination of a number of risks to the process.

- Capsules pre-sterilised by autoclaving or irradiation can be aseptically connected to other disposable or *hard piped* system components. Using pre-sterilised capsules eliminates the need for the user to sterilise the filter, removing the risk of filter damage during this treatment. The pre-sterilised item is immediately ready for use, saving time in preparation and assembly.

- The inability to steam sterilise a filter capsule in line means there is no risk of filter damage caused by uncontrolled steam in place systems. Furthermore, the filter element is protected inside the capsule from physical damage during handling. Taken together, the risk of a filter failing an integrity test prior to use is dramatically reduced, saving costs in replacing filters and investigation time.

- The filter element inside the capsule cannot be replaced — the entire assembly is single-use and disposable. The assembly vent/drain valves are not therefore repeatedly used and there is no risk of gaskets becoming damaged or leaking as a result of prolonged exposure to the process or sterilisation treatments.

- The filter assembly does not require cleaning, which means there is no risk of product carryover from one batch or product type to the next and cleaning validation is unnecessary for the filter. Personnel are also not exposed to cleaning agents and solvents required to clean the filter assembly.

- For the manufacture of aggressive, pathogenic or toxic products, the enclosed filter capsule offers good protection for process operators, who do not have to dismantle the filter assembly and dispose of the contaminated filter element.

- Plastic disposable filter capsules are not as heavy as their stainless steel counterparts, making them easier to handle, especially in confined or awkward places, such as barrier isolators. The reduced weight also poses less risk of injury or damage to personnel and equipment, respectively.

- The volume of capsule filters is commonly much less than for a steel filter assembly, so less space is required for the filter and there is less hold up in the filter of valuable product. This is increasingly important as new biopharmaceutical products are manufactured in small, high value volumes.

- Integrity tests of these capsule filters may be performed more reliably, with less risk of erroneous results due to leaks or fluctuating environmental conditions. The plastic outer casings conduct heat less readily and so are less affected by changes in temperature in the surrounding environment.

There are also some practical limitations to the use of filter capsules.

- The filter casings are quite rigid at typical ambient temperatures but become softer when heated. They may then be unsuitable for filtration of solutions at elevated temperature, since their safe maximum working pressure may be reduced. The casings are also affected by low temperatures, becoming brittle below the glass transition temperature of the plastic (–5°C for polypropylene, for example), limiting their use for refrigerated solvents or very cold liquids.

- Although quite rigid, the casings do allow some small movement. This is not normally important but can prevent a disposable filter capsule being used in applications where precise control of flow and volume is essential. Point-of-filling filtration is a case in point. A filter located between the pump and filling needle on a vial filling machine will experience pressure fluctuations during the cycle of the filling pump, while the needles create a resistance to flow. The filter capsule is therefore fractionally 'inflated' during the stroke of the pump, while it delivers the predetermined volume of liquid product via the needle(s) to the final container. At the end of the pump stroke, the shell of the filter capsule relaxes, as the pressure is relieved, pushing a small additional volume of liquid towards the needle(s), from which it will emerge as drips. By this time the filled vial is moving on to the sealing step and is replaced by an empty vial at the filling position; one or other of the vials (filled or empty) is therefore either contaminated externally by product solution, preventing subsequent satisfactory labelling or printing of the vial, or the fill volume is incorrect due to the additional drops of product falling from the needle. A small stainless steel filter housing and hard piping between the pump and filling needles are required to prevent *afterdrip*.

- An advantage of hard piped systems is that they can be steam sterilised *in situ* (steam-in-place). Capsule filters cannot generally be sterilised by in-line steam for safety reasons and those claiming to sterilise in this way have not achieved wide acceptance. Using capsules therefore requires making aseptic connections in a filtration system. Although laminar air flow work benches and barrier isolators improve the cleanliness of the operating environment, these connections still constitute a potential risk to an aseptic process.

Disposable Systems

The real revolution taking place in disposable equipment and materials for pharmaceutical applications concerns not just filters but entire processing systems. These are of particular interest to the biopharmaceutical sector, offering a number of significant benefits that extend beyond the immediate needs of the process itself. In addition to filters, disposable technologies are being applied to tubing, large volume bags for product mixing and sterile holding (for storage or transport), connectors and other components, even disposable bioreactors (Fox, 2005). The main advantages in these technologies are allied to flexibility in the manufacturing process, significantly reduced set-up and validation costs, and speed of response. A single manufacturing facility can be used for multiple product manufacture, using single use containers and equipment. This increases the number of clients that a contract manufacturing organisation can work with, and reduces the turnaround time for manufacture of preclinical and clinical batches. Meanwhile operating costs, plant downtime, capital investment and regulatory compliance requirements are dramatically reduced (Meyeroltmanns, 2005; Pora, 2006).

Cleaning and its validation in biopharmaceutical manufacturing require a lot of time and effort. Regulatory guidelines require appropriate procedures to ensure that cross-contamination and malfunctions do not affect product quality attributes (safety, identity, strength, purity) (FDA, 2004; EU, 2001). Manufacturers must document their cleaning validation procedures and results, which are singled out for review in guidance to inspectors (FDA, 2006). Eliminating these procedures reduces the regulatory burden to approval. The ideal system would be one of pre-sterilised modules or components that are ready to use as soon as they have been unpacked and interconnected.

Naturally, the implementation of disposable systems is not without its limitations. There are negative aspects to many of the advantages listed above.

- The introduction of many more disposable polymeric components into the process increases the potential for the product or process solutions to extract substances from the system. The range of substances can be restricted by careful selection of a minimum number of different materials; nonetheless compatibility and extractables of all materials must be confirmed.

- Biopharmaceutical processes frequently involve growth of mammalian cell cultures and some commercially-used cell lines have proved difficult to grow in disposable bag culture systems (Fox, 2005).

- There are increased variable and consumable costs per manufacturing batch, which, over time, may amount to a considerable sum of money.

- For dedicated production systems or large scale manufacturing (10,000 litres) it is difficult to justify a fully disposable system. Larger scale facilities may already have completed the necessary investment in vessels, fluid handling equipment and cleaning validation to enable multi-product use of the plant.

- The robustness of the disposable components is particularly important as scale increases, since the risks and implications of leaks become greater. The strength of polymeric components is reduced at elevated temperatures, making storage and handling of hot liquids more difficult.

- One benefit of small disposable systems is the possibility of moving components and systems during the process cycle. On a very large scale, moving bags full of liquid product may be difficult. This can be facilitated by the use of plastic liners inside mobile steel vessels, for example, but limits the potential benefits from the disposable concept.

- The maximum permitted pressure that can be used in a disposable system will be less that in a hard-piped system, especially when working at above normal ambient temperature.

The future of disposable systems is good, especially in contract manufacturing organisations where these technologies will find more and more applications at all steps of processing. Advantages from disposable systems outweigh potential disadvantages. On larger scales, where stand-alone bag systems will remain unsuitable, disposable bags can serve as liners for large steel vessels. These are less expensive capital items requiring neither electro-polished 316L stainless steel nor sanitary accessories. Indeed, conversion to disposable technologies is expected (Fox, 2005) to offer its major benefits in cost, quality and process planning. Some features of current traditional manufacturing operations, such as clean-in-place and steam-in-place, may even be avoided completely.

One objective of the biopharmaceutical sector is for specifically-targeted single-patient therapies. These would necessarily be manufactured in small volumes, making current disposable technologies ideally adapted to their needs. The demand for contract manufacturing services will increase, as production of pre-clinical, clinical and commercial product batches is outsourced from the developing organisation. In particular, virtual biotech companies with no facilities of their own, undertaking all development and manufacturing activities through contract services, will be able to set up the necessary manufacturing capacity they require in a very short time and in locations previously considered unsuitable.

REFERENCES

American Society for Testing and Materials (ASTM) (1983). *Standard test method for determining bacterial retention of membrane filters utilised for liquid filtration*. American Society for Testing and Materials F838–83, October.

Ball, P. (1999) Presentation at PDA international congress. Tokyo, Japan, 22–26 February.

Blosse, P.T. and Sundaram, S. (1999) Diminutive bacteria — implications for validating sterile filtration processes. *Pharmaceutical Technology Europe* **11** no 6, 20–26.

Booth, C. (2006) The role of media fills in process control. *Pharmaceutical Technology Europe* **18** (9), 40–44.

Cole, S.A. (1995) *Validating Filters for Aseptic Processes*. Presented at International Society of Pharmaceutical Engineering, March 1995

Cole, S.A. and Parker, A. (1985) Unpublished studies.

Detyna, J. (1995) Adsorption validation for aseptic filtration. *Journal of Validation Technology* **1** (4).

Docksey, S et al. (1999) A general approach to the validation of sterilising filtration used in aseptic processing. *European Journal of Parenteral Sciences* **4** (3), 95-101.

EMEA (1997) European Agency for the Evaluation of Medicinal Products (EMEA). Note for guidance: Manufacture of the finished dosage form (and Committee on Proprietary Medicinal Products).

European Commission (2001). European Commission working party on control of medicines and inspections qualification and validation. EU guide to Good Manufacturing Practice, volume 4, annex 15 *http://ec.europa.eu/enterprise/pharmaceuticals/eudralex/homv4.htm*

European Commission (2004) *EC Guide to Good Manufacturing Practice — Annex 1: Manufacture of Sterile Medicinal Products*.

FDA (1987) Guidance for Industry. *Guideline on Sterile Drug Products Produced by Aseptic Processing — Current Good Manufacturing Practice*. US Food & Drug Administration Pharmaceutical CGMPs. Center for Drugs and Biologics, Rockville MD, June 1987.

FDA (1993) FDA Title 21 of Code of Federal Regulations (CFR), parts 170–199.

FDA (1994(a)) Human Drug cGMP Notes. Center for Drug Evaluation and Research, Food and Drug Administration, Volume 2, No 3 (Sept 1994).

FDA (1994(b)) Human Drug cGMP Notes. Center for Drug Evaluation and Research, Food and Drug Administration, Volume 2, No 4 (Dec 1994).

FDA (2004(a)) FDA Guidance for Industry. *Sterile Drug Products Produced by Aseptic Processing — Current Good Manufacturing Practice*. US Food & Drug Administration Pharmaceutical CGMPs. Center for Drugs and Biologics, Rockville MD, September 2004.

FDA (2004(b)). Equipment cleaning and maintenance. Title 21 of Code of Federal Regulations (CFR), revision 25, part 211.67.

FDA (2006) FDA: Guide to inspections: Validation of cleaning processes (*www.fda.gov/ora/inspect_ref/igs/valid.html*).

Fox, S. (2005) Disposable BioProcessing. The impact of disposable bioreactors on the CMO industry. Biopharmaceutical Contract Manufacturing 2005: Improved processes and new capacity for pipeline to commercial production. HighTech Business Decisions, Moraga CA, USA.

Halls, N. (1995) Liquid media fills. Presented at International Society of Pharmaceutical Engineering, March 1995.

Howard, G. and Duberstein, R. (1980) A case of penetration of 0.2µm rated membrane filters by bacteria. *J Parent Drug Assoc* **34**, 93–102.

Leo, F. et al. (1997) Application of 0.1µm filtration for enhanced sterility assurance in pharmaceutical filling operations. *BFS News*, August.

Levchuk, J.W. (1994) Good validation practices: FDA issues. *PDA Journal of Pharmaceutical Science and Technology* **48** (5), 221–223.

Levy, R.V. et al. (1990) The matrix approach: Microbial retention testing of sterilising-grade filters with final parenteral products, Part 1. *Pharmaceutical Technology* **14** (9), 160–173.

Madsen, R. (1995) Filter validation. *PDA Letter*, June 1995, pp1, 10.

Martin, J. (2005) The future is disposable. *Pharmaceutical Technology Europe* **17** no 9, 66.

Martin, J. (2006) Private communication.

Martin, J. and Brantley, J. (1994) Parametric vs Matrix approaches to validation — clearing the confusion. *Filtration News*, Spring 1994 (Pall Corporation).

Meltzer, T.H. (1995) The significance of sieve retention to the filter validation process. *PDA Journal of Pharmaceutical Science and Technology* **49**, 188–191.

Meyeroltmanns, F et al. (2005) Disposable bioprocess components and single-use concepts for optimized process economy in biopharmaceutical production. *Supplement to BioProcess International* **3** no 9, 60–66.

Motzkau, P, and Okhio, L. (2005) The importance of vendor validation services: experience and economics. *Supplement to BioProcess International* **3** (9) 54–58.

Osumi, M., Yamada, N. and Toya, M. (1991) Bacterial retention mechanisms of membrane filters. *Pharmaceutical Technology Japan* **50**, 30–34.

Pall, D.B. and Kirnbauer, E.A. (1978) Bacterial removal prediction in membrane filters. Presented at 52nd Colloid and Surface Science Symposium, University of Tennessee, Knoxville, TE (June 1978).

PDA (1998). Technical Report No. 26, Sterilising Filtration of Liquids. *PDA Journal of Pharmaceutical Science and Technology* **52**, supplement no. S1.

Pora, H. (2006) Increasing bioprocessing efficiency — Single-use technologies. *Pharmaceutical Technology Europe* **18** no 1, 24–29.

Reif, O.W. et al. (1996) Analysis and evaluation of filter cartridge extractables for validation in pharmaceutical downstream processing. *PDA Journal of Pharmaceutical Science and Technology* **50** (6), 399–410.

Rohrschneider, L. (1973) Solvent characterisation by gas–liquid partition coefficients of selected solutes. *Analytical Chemistry* **45**, 1241–1247.

Sharp, J. (1995) What do we mean by 'sterility'? Commentary in *PDA Journal of Pharmaceutical Science and Technology* **49** (2), 90–92.

Snyder, L.R. (1974) Classification of the solvent properties of common liquids. *Journal of Chromatography* **92**, 223–230.

Snyder, L.R. (1978) Classification of the solvent properties of common liquids. *Journal of Chromatographic Science* **16**, 223–234.

Stone, T.E. et al. (1995) Methodology for analysis of filter extractables: a model stream approach. *Pharmaceutical Technology Europe* **7** (3), 27–34.

Sundaram, S. (1997) Presentation at PDA annual meeting, Philadelphia, PA, USA, 10–14 November.

USP (2002) United States Pharmacopoeia (USP25), National Formulary (NF20), United States Pharmacopoeial Convention, Inc., Rockville, MD.

USP (2005) *Biological Reactivity Tests, in vivo. <88>*. United States Pharmacopoeia (USP28), National Formulary (NF23).

Wällhausser, K.H. (1979) Is the removal of microorganisms by filtration really a sterilisation method? *J Parent Drug Assoc* **33**, 156–170.

Weitzmann, C. (1997) The use of model solvents for evaluating extractables from filters used to process pharmaceutical products. *Pharmaceutical Technology* **21**, 72–99; *BioPharm* **10** 4–5.

7

CLEANING AND PREPARATION

Nigel Halls and Stewart Green

INTRODUCTION

Numerous sources of regulatory and practical advice are available covering the critical aspects associated with sterile/clean pharmaceutical manufacturing, notably:

- Food & Drug Administration (FDA) have published aseptic processing guidelines since 1987 (subsequently updated in 2004 (FDA, 2004))

- the European Agency for the Evaluation of Medicinal Products (EMEA) has two Annexes attached to the GMPs, one covering sterile 'traditional' pharmaceutical manufacturing (Anon, 2004(a)) and the other biological preparations (Anon, 2004(b)).

The contents of these regulatory guidelines range from the highly prescriptive, eg the limits applicable to classification and monitoring of graded clean rooms, to interpretive, eg interventions to be included in media fill protocols. What is perhaps surprising, given the potential impact on the product, is that little guidance is given on how to prepare and subsequently store items used within aseptic manufacturing processes prior to their sterilisation/sanitisation. These items range from direct product contact items such as stoppers (closures) to indirect contact items such as engineering tools, cleaning equipment, etc.

In the authors' experience most manufacturers of sterile pharmaceutical products consistently meet inspectional requirements for process validation, equipment validation, and for equipment control and maintenance, ie citations against biological and thermal qualification of autoclaves, monitoring and control of pressure differentials, and against personnel training rarely feature in inspection reports. However, failure to effectively sterilise difficult items such as hoses and pumps, or to appropriately store such items, still features in many inspection reports.

This chapter will explore the cleaning and preparation of these ancillary items, hopefully providing sufficient guidance to ensure that they should not elicit adverse regulatory comment and more importantly will not compromise the sterility of the products with which they are associated.

Regulatory Guidance

There is very little regulatory guidance on how to prepare materials for sterilisation. Guidance from FDA (FDA, 2004) and from the EU (Anon, 2004(a)) is summarised in Table 7.1 (pp248–249).

Where there is guidance (see Table 7.1), it can be seen that there is significant commonality in the guidelines in terms of their general approach which can be summed up as 'clean appropriately', 'protect appropriately after cleaning', 'sterilise appropriately', 'store only for a defined period and then under appropriate conditions'. There are practically no prescriptive regulations nor recommendations. For instance, (perhaps not unexpectedly), there is no specific advice other than in Annex 1 (Anon, 2004(a) concerning the classification/grading of the areas in the factory used for these preparatory processes. The lack of prescription is because different types of item used by various manufacturers of aseptic sterile pharmaceuticals may be subject to different sterilisation processes, may be presented in different configurations and may be used under different conditions, thereby posing a range of different challenges too great to lend itself to prescription. For this reason most sterile product manufacturers have had to evolve their preparation and cleaning methods based on a combination of validation, experience and, regrettably, problems identified at inspection. The remainder of this chapter will explore this subject and try to provide more specific guidance using the authors' experience and an appreciation of the risks which may be encountered.

Scope

This chapter will explore the following items for sterilisation:

- non product contact consumables

- product related contact items eg stoppers
- product contact but not product related equipment eg pump parts

... and address their sterilisation by:

- dry heat
- moist heat
- irradiation
- vapourised hydrogen peroxide (VHP), often used in conjunction with isolators.

In looking at each specific item and sterilisation method the authors will attempt to identify the rationale behind each step involved in the process, from cleaning to use. While clearly it is not possible to consider every item that may be utilised by different manufacturers, the principles involved should for all practical purposes be the same.

CLEANING

All sorts of things require cleaning prior to their use in aseptic sterile manufacture.

For instance, some items of equipment may be re-usable. In these cases it is necessary to prevent adulteration or cross-contamination of subsequent products by ensuring that traces of previous products (or even traces of degradants of the same product if run previously) are effectively removed. Most effort is applied to removal of molecular traces of active pharmaceutical ingredients (API), but excipients, preservatives, placebos and even cleaning agents themselves must also be cleaned from the product-contact surfaces of re-usable items of equipment. The regulatory agencies have taken a considerable interest in cleaning and 'cleaning validation' since the early 1990s (FDA, 1993; Anon, 2004(c)). This interest was stimulated by an incident in which a finished drug product (Cholestyramine Resin USP) was recalled in 1988 due to contamination by traces of an agricultural pesticide. Drums used to store the pesticide had also been re-used for storing solvents involved in the manufacture of the active pharmaceutical ingredient.

In sterile manufacture it is also common (but not universally so) to clean new supplier-sourced single-use product contact items, such as rubber stoppers, glass

Table 7.1 Regulatory Guidance to Preparation of Items Prior to Sterilisation

General Requirement Concept	EU GMP Annex 1 (Anon, 2004(a))	FDA, 2004 (FDA, 2004)
'Clean appropriately'	A1.2 *The various operations of component preparation ... should be carried out in separate areas within the clean area* A1.11 *Preparation of components* and most products (for terminal sterilisation) should be done in at least a grade D environment in order to give low risk of microbial and particulate contamination ...*	
'Protect appropriately after cleaning;	A1.49 *Components*, containers and equipment should be* handled *after the final cleaning process in such a way that they are not re-contaminated.* A1.50 *The interval between the washing and drying and the sterilisation of components* ... should be minimised and subject to a time-limit appropriate to the storage conditions.* A1.12 *Components* after washing should be handled in at least a grade D environment.*	VI.B.1 *The time between washing, drying and sterilisation should be minimised because residual moisture on the stoppers can support microbial growth and the generation of endotoxins.*

General Requirement Concept	EU GMP Annex 1 (Anon, 2004)	FDA, 2004 (Anon, 2004)
'Sterilise appropriately'	A1.53 *Components*, containers, equipment and any other article required in a clean area where aseptic work takes place should be sterilised ...* A1.67 *Items to be sterilised ... should be wrapped in a material which allows removal of air and penetration of steam but which prevents re-contamination after sterilisation*	
'Store only for a defined period and then under appropriate conditions'	A1.50 *The interval between ... sterilisation (of components) and use should be minimised and subject to a time-limit appropriate to the storage conditions.* A1.60 *Each basket, tray or other carrier of products or components* should be clearly labelled ... and an indication of whether or not it has been sterilised.*	

- the word 'component' is used differently in EU and FDA documentation.
  FDA use the word 'component' to describe starting materials, ie components within a formulation or dosage form.
  EU GMPs use the word 'component' to describe packaging materials, ie components contributing to the final presentation.

bottles, vials ampoules, etc. This has more to do with taking responsibility for the quality and contamination potential which could arise from items which may, at least in principle, have been contaminated by chemicals used in their production. Alternatively, they could have been contaminated physically by particles, dirt, soil, etc while in storage or transit between the supplier and the pharmaceutical manufacturer.

For example rubber stoppers/closures used to seal vials may be 'contaminated' by release agents (commonly silicones or stearates) used to facilitate release of the stopper from its mould. Most stopper manufacturers (it has to be said) pre-wash their stoppers to remove, as far as possible, this 'contamination'. Nevertheless, during product development, stopper studies are normally performed to ensure that the drug product is not compromised by the stopper ingredients intentionally or unintentionally. Sensitive biological products have had their potency significantly reduced by interaction with stopper components. In addition to these molecular contaminants, rubber stoppers may also be physically contaminated by particulate dirt (compression moulding is hot and humid work) and by pieces of broken-off rubber. It is not unusual for the particulate contamination on rubber stoppers to increase during shipment, due to the abrasive effects of the stoppers moving against one another as a result of the inevitable buffeting they receive in transportation.

The unique aspect of cleaning, as far as items of equipment and product-contact components are concerned, in relation to sterile manufacture, is that the cleaning processes are not only expected to remove molecular contaminants and physical particles, they are also expected to reduce bioburden (in the meaning of reducing the numbers of microorganisms contaminating the items).

The cleaning method used must meet two essential requirements:

1. It must effectively remove the contaminants (from whatever source) — this may include micro-organisms (see below) — by using cleaning agents and cleaning equipment which are neither damaging nor corrosive to the item being cleaned.

2. The cleaning agents themselves should be soluble and free rinsing so as not to leave residues which in themselves may compromise the product.

It should not be assumed that items being cleaned are totally inert or that they all present perfectly smooth impervious surfaces. Stainless steel, even if micro-polished, does not present a perfectly smooth surface in terms of residue adherence even though it may visually appear smooth. Pipework or pump nozzles made from extruded tubing, when microscopically examined even under low magnification, can usually be seen to be striated/scored along the length of the tube. Micro-polishing removes the peaks of the striations, sometimes by folding them into the troughs, but never provides a completely crevice free surface. The same is true of 'plastic' tubing made from extruded polymeric materials. It is, however, debatable as to just how the extent of surface roughness (or conversely smoothness) may affect adhesion of microorganisms to surfaces and the difficulty involved in their removal. There is evidence (Riedewald, 2006) to suggest that the use of highly polished surfaces to minimise microbiological attachment and to maximise cleanability is a misconception. This of course does not mean that the quality of surface finishes can be ignored, only that extreme measures to make them extra smooth may be unnecessary.

Ideally, cleaning would involve only water, normally *Purified Water Ph Eur/USP*, finishing (in the case of items used in connection with the manufacture of sterile parenteral products) with *Water for Injection(s) PhEur/USP* as the final rinse. Where contamination is difficult to remove, the first option is to supplement water with some mechanical means of removing it — water under pressure is the most straightforward application of this concept, abrasive pads and brushes are more complicated ones.

In fact many manufacturers do not consider the use of hot *Water for Injection(s)* but immediately opt for chemical cleaning agents throughout all their cleaning applications. Nonetheless, this option should not be taken without some considerable deliberation regarding its necessity. In the eyes of the regulatory agencies (at least) it increases the complexity of the cleaning process along with their expectations of the validation undertaken. This in turn leads to a greater risk or unsatisfactory inspection outcome. If the product is free rinsing, non-staining and does not contain appreciable amount of fats or proteins then *Water for Injection(s) PhEur/USP* with, where necessary, mechanical support, may often suffice. Even difficult to clean items may be amenable to water-cleaning if allowed to soak in hot *Purified Water* or *Water for Injection(s)* for a defined period prior to cleaning.

Where contaminants are not soluble in water or have a tendency to bind to the surface, then there can be no choice but to consider the use of chemical cleaning agents. The first choice is normally use of detergents, usually obtained from proprietary sources. These aid cleaning by various means. First, they increase the 'wettability' of surfaces by alignment of surfactant molecules at the boundary between the contaminant and the water used for dissolution. Second, by emulsification, in which the contaminants are broken up and suspended in the cleaning water as emulsion droplets. Third, by dispersion, in which the emulsion droplets are retained in suspension so they do not aggregate and get re-deposited on the surfaces being cleaned.. Many proprietary detergents may also contain chelating agents, and acids or alkalis to assist in the removal of proteins and fats respectively. For a proprietary detergent to be approved for use in the pharmaceutical manufacturing industry, the supplier must be able to provide a full validation service for the industry, either directly or through its agents. This should include:

- compatibility data against materials commonly used in pharmaceutical manufacture
- toxicity data
- effectiveness data
- technical support

Technical support should include the carrying out of trials on behalf of the pharmaceutical manufacturer to demonstrate cleaning effectiveness, and particularly for providing analytical methods for residue detection. It is also important to ensure that the detergent supplier agrees to alert the user to any significant changes in the formulation, manufacture or packaging of its cleaning agents. Changes in concentration, substitution of the surfactant or cleaning agents, changes in stability, new evidence of compatibility/toxicity etc, should all be included in a formal technical agreement. This is quite different from the commercial agreement which deals with prices, quantities to be purchased, order and delivery conditions, etc, and should be specifically negotiated prior to approval of a supplier of proprietary detergents. Any changes of the types described above may well invalidate previous cleaning residue studies with potentially serious effects on a pharmaceutical product, for example, deleterious impacts on stability leading to recall or worse.

Removal of micro-organisms and their cell residues (in particular endotoxins in the case of sterile parenteral products) are not likely to pose greater problems of removal than those applicable to chemical residues. It has been known since the early 20th century (Hamilton, 1931; du Moulin et al, 1981) that some microorganisms, particularly if they proliferate, can produce intractable 'gummy' exudates, particularly resistant to cleaning. However these situations have a very low probability of occurrence in pharmaceutical manufacture.

Endotoxins are normally readily removed by water cleaning alone or by water plus surfactant.

There are broadly two approaches to cleaning of equipment — manual cleaning and 'clean-in-place' (CIP) systems. In the former, the equipment is stripped down and taken to a cleaning area (or taken to a cleaning area and then stripped down). In CIP, the equipment remains intact and is cleaned by (usually) a combination of installed water jets, water turbulence, and rinsing. CIP may include detergents.

Manual cleaning is potentially subject to greater variability than CIP. The advantage here is that the equipment is stripped down and therefore (assuming that the work has been defined properly, the cleaning operator has been trained and is supervised) the difficult to clean locations are known and can be addressed. Of course they are not always addressed wisely — abrasive pads can sometimes do more harm than good, bristles released from over-used brushes rather negate the investment in (say) high efficiency particulate air (HEPA) filtered air protection provided, as it sometimes is, to the wash area. Where manual processes are used 'Murphy's Law' applies — 'if it can go wrong — it will go wrong'. Cleaning areas and materials used for manual cleaning need to be very carefully specified, controlled and supervised. Sometimes the simplest tasks are the hardest.

Initial stages of manual cleaning are most often done using potable quality hot tap water. Final stages must always be carried out using the appropriate pharmacopoeial grade water for the pharmaceutical product which will subsequently come into contact with the cleaned equipment (EMEA, 2002).

There is no 'generic' CIP system. Each differs as they should be 'tailor made' for each particular application. They may use water alone (hot or cold, or both hot and cold) or they may use water plus detergents, etc. As with manual cleaning, the final rinse should be with the appropriate pharmacopeial grade of water. Many CIP systems are computer controlled, and therefore have much less variability than manual cleaning. Some, however, are manually controlled and the number, sequence and timing of various operations with valves (some of which may be relatively poorly accessible) may be very complex — so it should not necessarily be assumed that CIP is always 'better' than manual cleaning. In the long run CIP systems can only perform as well as their design, and design flaws in CIP systems are exceedingly difficult to rectify. Valves and pumps are not stripped down, therefore the spray balls, water jets, drains, etc, involved in their make-up must have been correctly placed in the first instance and remain functioning in the manner intended through repeated applications.

Cleaning Validation

It is not intended in this chapter to provide a full exploration of cleaning validation. The underlying idea is that pharmaceutical manufacturers should be confident that when they come to use a piece of equipment for manufacture of one product that it has been satisfactorily cleaned, in the sense that it does not carry residues of the previous product for which it was used. The reader is recommended to consult publications by FDA (FDA, 1993), the Pharmaceutical Quality Group of the UK Institute of Quality Assurance (Anon, 1999) and the excellent web site *http://cleaningvalidation.com* and publication by Destin Le Blanc (Le Blanc, 2005(a),(b)).

All regulatory authorities now expect to see validation evidence that either the product (or its degradants) and/or any cleaning agent used has been effectively removed from product contact surfaces of production equipment, particularly for equipment used for multiple types of product. The standard approach is to use the toxicity data for either the product or the cleaning agent, and then build in a safety factor for the acceptable residue (commonly 1/1000th or 1/10000th of the LD_{50}). Calculate the surface area of the product contact surface and then calculate the allowable residue using the factorised toxicity data and the surface area. Exactly the same approach can be taken with the cleaning agent.

But what of removal of micro-organisms? It is doubtful that any specific cleaning measures need to be taken in excess of those used to remove chemical residues — generally it would seem reasonable to suppose that cleaning processes which are capable of removing contaminants or residues of molecular dimensions should be more than capable of removing contaminants of physical dimensions, such as microorganisms. As stated previously in this chapter (see p251), cleaning with either hot *Water for Injection(s)* or surfactants and alkaline or acid cleaning agents is normally sufficient to meet the criteria applied to chemical residues.

FDA (FDA, 1993) stresses that microbiological cleaning validation is not a matter of setting acceptance criteria and testing against them, but one of process control prior to the use of equipment. Microbiological challenges on equipment and product-contact components requiring cleaning need never be unduly high. Although many many different scenarios could be considered, the main ones are as follows:

1. Product-contact components must be obtained from approved suppliers. One of the functions of supplier audit is to ensure that reasonable standards of hygiene are maintained on their premises. Sometimes the suppliers' processes are intrinsically anti-microbial. For instance, glass may be exposed to temperatures of more than 800°C in its manufacture — unless there is some gross mis-management of glass after manufacture and prior to receipt by the pharmaceutical manufacturer, glass components are most unlikely to carry a heavy bioburden.

2. Product-contact equipment, in the way that it is used in connection with most sterile pharmaceutical products, is likely to have been repeatedly flushed through with materials which carry no bioburden, and in many cases which are themselves anti-microbial or preserved. The risk of microbiological contamination of such equipment lies in the length of the interval and the conditions in which it is left after its use and prior to cleaning. Completely closed systems with CIP provide the best control over the conditions between use and cleaning, but the time interval could still be consequential.

3. In some processes it is known that equipment may be exposed to viable microorganisms, eg in fermentation and in manufacture of some biologics. In these cases it makes sense, not only in terms of good manufacturing practice (GMP), but also in terms of health and safety to decontaminate such materials either at high temperatures or by exposure to chemical disinfectants before cleaning. The bioburden for cleaning should therefore be low.

The real issue in relation to validation of cleaning processes versus removal of microorganisms is minimising the challenge by appropriate controls and process steps. This is best done by a thorough appraisal of the factors in the process which could allow (or conversely prevent) proliferation of microorganisms on the equipment, and identification of any vulnerabilities in its design or construction which could adversely affect the cleaning process. Based on this it is difficult to see that setting specific guidelines for microbiological cleaning validation could be justifiable.

Cleaning after Media Fills

In the context of the overall theme of contamination control within this book, there are issues to be addressed around the cleaning of microbiological growth media from pharmaceutical manufacturing equipment exposed during media fills (broth trials, simulation trials, etc).

Generally speaking, a growth medium such as Soybean Casein Digest Broth (SCDB) resembles a pharmaceutical product comprising dissolved ingredients, which are themselves products of chemical synthesis only in that both are aqueous liquids (Halls, 2004). In other respects they differ. SCDB was designed to encourage and support the growth of a wide range of microorganisms: this is not one of the objectives of the formulation scientists involved in designing new injectable products. In fact they often have the objective that the formulation should do exactly the opposite — namely to prevent or inhibit the growth of microorganisms.

This creates a wholly different cleaning problem — very small traces of chemical residues within the allowable acceptance criteria left on equipment

Figure 7.1 Microbiological Contamination in Media FIlls Arising from Traces of Residual Media in 'Difficult-to-Clean' Locations

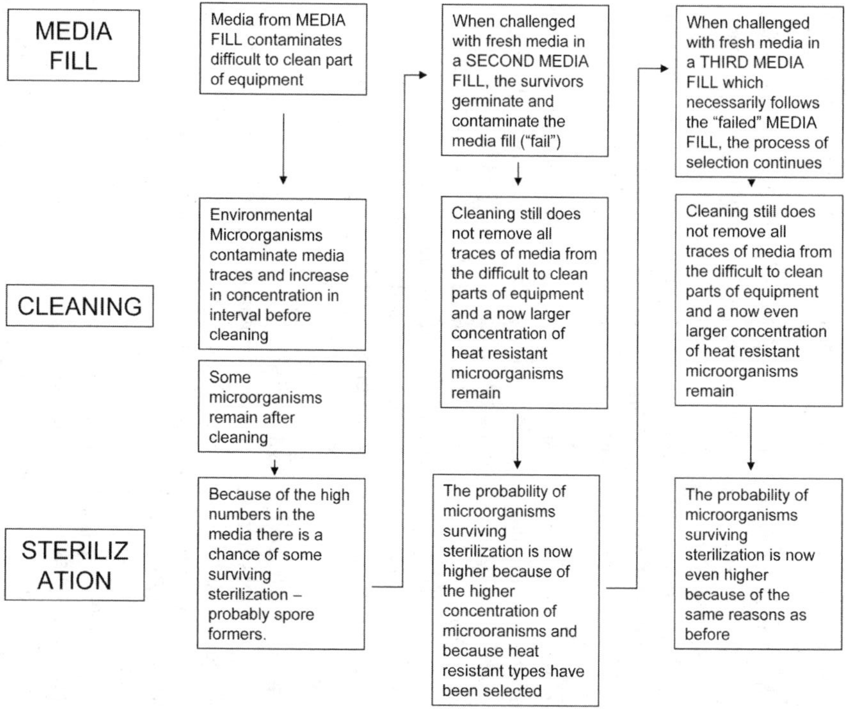

remain as very small traces, they do not increase to larger and larger traces. The same very small traces of microbiological growth media on equipment may result in very minor incidences of microbiological contamination leading to unsatisfactory and problematic levels.

For instance *x* µg of (say) Ranitidine Hydrochloride remaining beneath a (say) valve seating remains as *x* mg of Ranitidine Hydrochloride. If contaminated by one environmental microorganism from a cleaning operator's skin flake, the microorganism is most likely going to die. On the other hand, if that same one microorganism were to contaminate a trace of *x* mg of SCDB beneath the same valve seating, there could (at least in theory) be about 1,000 microorganisms in around six hours after the contamination event.

This actually happens — frequently in complex equipment, most often in sterile API manufacturing equipment subject to CIP cleaning systems. Even though it is not mandatory to use microbiological growth media for simulation in this type of equipment (PDA, 1996), many companies have tried it and experienced validation delays as a result.

Figure 7.1 summarises a pattern of contamination in media fills, most often seen when consecutive media fills are required, as in initial validation of new equipment or following a media fill 'failure', to confirm satisfactory corrective action. In an example seen by one of the authors of this chapter there was media contamination between a pin-hole damaged diaphragm on a Saunders valve and its seal, but these traces of media were only confirmed after the problem had arisen and after the valve (and many other valves, etc) had been stripped down. The first media fill was successful.

The traces of media left in the valve, it was then surmised, became contaminated with whatever microorganisms may have been present in the immediate environment, they increased in number and remained after cleaning. Some survived sterilisation because of the higher than normal numbers. Later events indicated that this was most likely true because contaminants in subsequent media fills were heat resistant spores of the genus *Bacillus*.

As consecutive media fills were undertaken, and because there was no apparent problem with the first media fill, the second was performed. It failed due to contamination with *Bacillus* spp. The surmise was then that this was because surviving spores were able to germinate when coming into contact with the fresh SCDB which was run into the equipment.

Cleaning and sterilisation were carried out immediately. The problem was investigated, but no root cause was determined (as is very common with microbiological problems). When every conceivable minor issue had been addressed, a third media fill was performed. It failed more extensively than the second, still *Bacillus*. It was surmised (after the event, naturally) that the processes of cleaning and sterilisation were 'selecting' this particular microorganism, whose numbers were increasing behind the diaphragm every time new media was allowed to 'refresh' its supply of nutrients.

The equipment was completely stripped down and replaced where necessary. A different method was used for simulation thereafter, without using SCDB in the equipment. It should be emphasised that this was a problem purely associated with the use of media. Had the product being run through this piece of equipment penetrated behind the diaphragm there would have been no issues, because there was conclusive evidence that microorganisms were not capable of growth in this product.

In summary, media fills may create problems of their own making. The decontaminating power of pharmaceutical products should not be underestimated, but should not of course be relied upon. When performing media fills, a thorough assessment of the microbiological risks should be undertaken (and at the very least, for regulatory purposes, documented). It may be wise to perform both two successive cleanings and sterilisations on equipment exposed to media even when there is validation evidence showing only minimal traces remaining. This invites the question — why do the same thing twice, if it does not work the first time, why should a second time be any better? The answer is that both rinsing and sterilisation processes function exponentially — there can be no guarantee that everything is removed or that everything is killed, but with each treatment the probability gets lower.

PROTECTION PRIOR TO, AND DURING, STERILISATION

Following the cleaning process it is essential that all items are adequately protected to prevent re-contamination. This requirement is applicable to chemical contamination (materials left 'lying around' after cleaning may be re-soiled at the molecular level by other detergents or chemicals used for other purposes), to physical contamination (particles), but particularly to microbiological contamination. Chemical and physical particles have no capability to reproduce themselves — but contamination by a single microorganism on one day could lead to contamination by several thousands or even millions some time later.

The levels of protection used vary from manufacturer to manufacturer and from item to item. It is possible that cleaned items may have minimal protection — such as storage on open shelves in a 'clean' environment (probably best supplied with HEPA filter air). On the other hand, it is more convenient (and in truth better GMP because it minimises manipulations of previously cleaned equipment, etc) to protect and store cleaned equipment in containers or wrappings suitable for the sterilisation (which in the case of manufacture of sterile products is the next stage in the process). For this reason protection may be provided by storage in metal boxes or 'drums' or in flexible packaging (bags or wrappings) ready for sterilisation. Bags are available in Tyvek® or specific grades of paper (see below). Whichever method is chosen it is essential that items are clearly marked:

- with their status, ie cleaned
- with the date of cleaning
- with the expiry date or the date when an item must be either 'refreshed' (which usually amounts to rinsing with *Water for Injection(s)* or re-cleaning using the validated process).

The regulatory agencies expect expiry dates to be established by specific studies. Such studies are often 'token' evaluations carried out to support some rational, reasonable or sensible storage period that meets production needs, often they are in the order of five days before refreshing and 10 days before re-cleaning. Intact physical barriers are not penetrated by microorganisms in clean indoor environments unless subject to undue stresses or when materials are left wet.

It is essential that all items, particularly those which may not be free draining, eg tubing, stoppers, etc, must be stored dry. Although comparatively little research has been undertaken on microbial growth rates in/on 'natural' pharmaceutical manufacturing environments (Shimomora et al, 2006) the perceived commonsense wisdom is that any significant moisture on cleaned items will encourage microbial proliferation which may compromise subsequent sterility of the item or its freedom from pyrogens. Most microorganisms capable of proliferation in low nutrient environments, typical of those which might be expected on cleaned equipment etc, are gram-negative. These are generally highly sensitive to most sterilisation processes, and are, on the other hand, the most prolific generators of endotoxin (pyrogen). Removal of endotoxin cannot be guaranteed by any of the commonly used sterilisation processes, except dry heat.

Dry Heat Sterilisation

Items for dry heat sterilisation are generally product-contact items such as vials, pipes, tools etc. The main requirement is that they are unaffected by temperatures in the region of 250°C for exposure periods in excess of one hour. Not only do they need to be thermally stable but critical attributes must not be impacted. For example, pump parts for an interference fit pump are not normally dry heat sterilised in case 'fit' problems arise due to different amounts of expansion between the pump chamber and the piston (usually mechanically 'matched').

Equipment for dry heat sterilisation in ovens is generally presented in closed metal containers (mainly stainless steel, but occasionally aluminium, although this material is more likely to generate particles which could lead to product rejection). These items are subsequently stored in the same container having been removed post-sterilisation into either the sterile core (Grade A/Class 100) or into a storage or cooling area of a lesser grade. The second option is relatively unusual in modern designs for aseptic facilities as it presents the problem of moving the components into the Grade A/Class 100 area without re-contaminating the item, or bringing contamination into the higher quality area (see below).

Glass vials for dry heat sterilisation may in small facilities be sterilised using dry heat ovens. In larger scale operations sterilisation (and depyrogenation) is more commonly done using continuous tunnel dry heat sterilisers. Temperatures of up to 350°C are achieved. Generally tunnels are part of an on-demand continuous (or semi-continuous) process, in which sterilisation is preceded by washing, and followed by filling. The three processes have to be balanced.

1. Washers are usually linked directly to sterilisation tunnels with no physical separation. The vials can therefore be loaded directly onto the washers in a pharmaceutically clean area (say Grade D which has no comparable US class, probably because its specification is so weak) and do not require any protection after washing. Automated washers start by inverting the vials which then remain upside down to allow any dirt or particles to drain out of them though multiple washes. The first stage may be hot tap water of potable quality or the same mixed with re-cycled *Water for Injection(s)* from the final rinse stage of the process. After the final rinse they are air-blown dry and fed into the tunnel — there may be an accumulator turntable to buffer the supply.

2. There are two main technologies for raising the temperature of vials in dry heat tunnels — by radiant heat elements and by unidirectional heated air. Tunnels of the latter type tend to be more compact. The rate of vial throughput on a continuously moving belt is controlled to ensure a specified

peak temperature is achieved for all vials. After heating in dry heat tunnels, vials have to be cooled to a temperature at which they can be filled without causing product deterioration or degradation. The final or output zone of the tunnel is given over to cooling by flushing with cold unidirectional HEPA filtered air — the so-called cooling zone. There is always an accumulator turntable between tunnels and filling lines, but here it has two functions — first to provide a buffer stock of vials, and second, because tunnels should not be allowed to stop unnecessarily, and definitely not with the frequency at which most filling lines stop for adjustment, etc.

Where vials are sterilised in dry heat ovens they must be separately washed. The principles of washing — inverting, finally rinsing with *Water for Injection(s)*, and drying with air, are the same as in continuous washers. However the vials must be protected from re-contamination after washing and before oven sterilisation. This is usually done by loading them into metal boxes or hybrid tray-boxes, and is most often carried out under unidirectional HEPA filtered air protection.

Dry heat processes are used not only to sterilise but perhaps more importantly to depyrogenate (ie to remove, inactivate or reduce bacterial endotoxins) items for use in connection with manufacture of sterile parenteral products. In the authors' experience most items used within the production of sterile parenterals do not present a significant 'pyroburden'. In studies performed by one of the authors glass vials after the washing process alone showed no evidence of endotoxins and the washing procedure itself was sufficient to achieve the 3 log endotoxin reduction expected for regulatory compliance.

Steam Sterilisation

Steam is the sterilisation process of first choice for items of equipment which can withstand temperatures in the range of 110°C to 135°C. It is also frequently used for equipment which can withstand the much higher dry heat temperatures because sterility can be achieved in significantly shorter cycles, thus providing far greater through-put capacity for the equipment.

Steam sterilisation is generally done in autoclaves, although some major items of equipment may be sterilised-in-place (SIP) using steam under pressure. Although continuously operating moist heat sterilisers are in use in the food canning industry, moist heat sterilisation of equipment in pharmaceutical manufacture is always operated as a batch process using steam.

There are a variety of 'systems' used to protect equipment prior to and during steam sterilisation. The fundamental requirements for containers and wrapping materials used to protect equipment items before sterilisation by steam and during sterilisation by steam are as follows:

1. The materials of composition and the design of containment systems must be such that they are barriers to microbiological ingress.

2. The materials of composition and the design of the containment system must be such that they allow passage of air (outwards) and steam (inwards). This is because in steam sterilisation the transfer of heat energy necessary to kill microorganisms is achieved by condensation of the steam on their surfaces. This is not effective if air is present to insulate the microorganisms from contact with the steam.

3. The design of containment systems must allow for them sufficient robustness to maintain their microbial barrier properties during handling, storage and sterilisation, while at the same time allowing accessibility of the equipment for use without recourse to undue force, and without generation of particles which could come to contaminate product, or even the equipment itself.

One option is to use metal containers. Microorganisms cannot for sure pass though a sheet of stainless steel, nor, however, can air or steam. To allow movement of gases and vapours, metal containers are generally equipped with perforated side sections which, prior to and after sterilisation, are closed by means of a movable metal plate. During sterilisation, the plate is shifted so that the perforations are open. Metal containers are also equipped with perforated bottoms to ensure that any condensate formed in the sterilisation process drains away — it is common sense that water condenses when hot steam hits cold stainless steel. When considered bleakly, metal containers with perforated sliding parts could be perceived as poor barriers to ingress of microorganisms, and, unquestionably, circumstances of pressure differentials, 'liquid bridges' etc, can be visualised in which microorganisms could gain access and contaminate the contents. Conditions of storage after sterilisation need serious consideration, unidirectional HEPA filtered air storage is probably the safest option (see below).

Alternative systems involving metal containers include perforated mesh lids protected by bacteria-retentive membranes which permit steam penetration but provide effective barriers to microorganisms. These type of systems are most often used for small items which can be difficult to place in autoclaves or may be damaged if moved around the autoclave by incoming steam pressure.

By far the most common means of protecting items for steam sterilisation is flexible packaging. There has been (surprisingly perhaps) very little published work or standardisation of these flexible materials for pharmaceutical applications, compared with the interest from hospitals (*http://www.aorn.org/proposed/PackagingSystem.pdf*) and from the sterile medical devices industry (ISO, 1997; Anon, 1997).

Irrespective of the published work, all three industries — hospitals, sterile medical devices and pharmaceutical manufacture — use either 'medical grade' paper or spun bonded poly-olefines (Tyvek®) for wrapping materials for steam sterilisation. Both of these materials are permeable to passage of small molecules (the gases making up air, steam, etc) and are reasonably durable. Neither paper nor Tyvek® provide an absolute barrier to passage of microorganisms (Sinclair and Tallentire, 1981; Tallentire and Sinclair, 1985), but both types of material appear to act as some form of fibrous filter by trapping microorganisms in their interstices. Except where there are significant pressure differentials and challenges, they provide adequate barriers to microorganisms.

These materials are available to the pharmaceutical industry in various configurations. In the simplest form they come in reels for use as wrapping. They are also available as reels bonded on the two edges to transparent plastic films — the reel stock can be cut to whatever lengths the end user wishes and heat sealed to form bags. They can be purchased pre-formed into bags (either with both sides made of the chosen material, or with one side of the chosen material and the other side made of a transparent film) — these, of course, are available in various sizes. The advantage of the transparent side to reel stock or to bags is that the contents can be seen. Although there are regulatory requirements for sterile equipment containers to be comprehensively labelled with the identity of their contents, it is undoubtedly helpful to the end user to see what he is about to use. Re-sealing sterile containers opened in error, although not specifically stated in any regulation known the authors, should not be allowed. Where flexible wrappings made from two materials are used for steam sterilisation, it is important that the presentation of materials in the autoclave does not result in occlusion of the air and steam permeable material.

Suppliers of flexible packaging (*http://www.healthmark.info/Steriking/Info&testData.pdf*) for sterilisation are generally helpful in providing technical details of their materials and pointing out advantages over competitor products.

Most technical attention has always focused on the microbiological barrier properties of flexible wrapping materials themselves. The authors are unaware of any published detailed scientific work concerning the sealing of these materials, either to themselves or to other films versus microbiological penetration. Perhaps this is unnecessary anyway. There are various methods of sealing — proprietary bags or reel stock are always multiply factory-sealed either by heat or by adhesive bonding. Seals placed by the pharmaceutical manufacturer should either be double heat seals, or the wrapping should be folded over at least twice and thoroughly taped down. Heat sensitive indicator tape is usually used for this purpose, giving the additional perceived advantage of distinguishing material that has been autoclaved from material which has not.

One final factor to consider is ensuring the consistent placement of wrapped items within the autoclave. In many cases this does not pose a problem but in the authors' experience some items, particularly lengths of tubing, can, if orientated differently, pose very different challenges to steam penetration. For this reason some manufacturers will coil tubes to a specific diameter before placing the item in its wrap.

Radiation Sterilisation

Radiation sterilisation normally, but not invariably, by exposure to 25 kGy from a Cobalt 60 source poses problems for the preparation of items which are different to those of high temperature methods of sterilisation. Irradiation can significantly change items exposed to effective sterilisation doses. For example, plastics can darken, become brittle, or lose elasticity. Migration of the plasticisers may also occur. Fabric may break down and paper can darken and become more fragile and prone to puncture.

Some suppliers of double-wrapped ready-to-fill sterile syringes use irradiation for the sterilisation of their barrels, almost all use irradiation for sterilisation of the rubber syringe plunger tips.

Clean room clothing is also frequently irradiated. In this case it is essential to carefully consider the fabric, as some fabric types rapidly degrade and will either tear or start to shed large numbers of particles ('linting'). Ceramic Terylene® is often chosen for its relative resistance to irradiation. However, many manufacturers or the service providers they use, do set limits on the number of cycles to which the garments are exposed, even for Ceramic Terylene®. Colour fading is an easily recognisable problem, deterioration of stitching thread may be more difficult to discern until after seams have split. Plastic zip fasteners are unsuitable for radiation. Common limits are 15–20 cycles, with some companies allowing up to 50 cycles, after which garments are replaced, but this depends not only on science, but on risk versus economic factors, eg usage rates.

Microbiological barrier properties are important for whatever material is used to protect items undergoing irradiation from recontamination by microorganisms. With radiation (in contrast to steam sterilisation) however, there is no requirement for molecular movements across the barrier material — protective materials can therefore be completely impermeable. Gamma radiation penetrates all materials. The extent to which its penetration is blocked is a function of the density and the thickness of the material through which it is passing — this is the principle of 'shielding'. The ability of materials to 'shield' or attenuate gamma radiation is illustrated by their Half-Value Layers (HVL) — the thickness of the material which reduces the intensity of irradiation to half of its 'entry' value. The HVL for lead is 12mm, for steel 21mm, and for water 200mm. It is evident from these HVLs that all practical flexible wrapping or even metal containment materials used for items subject to radiation sterilisation present negligible barriers to radiation penetration.

Of greater importance than the material used for direct protection of items destined for radiation sterilisation, is the consistency of the presentation of the items to the radiation source. This is in part dealt with by the design and construction of the irradiator in which some carrier mechanism or container which fits a conveyor system is used to expose the materials for sterilisation to the source of radiation. The product presentation delivered by the client must fit into these carriers or containers.

Most irradiation is done by specialist contractors to whom the client delivers items for irradiation in transportation containers, which irradiation contractors prefer not to have to open or tamper with.

Irradiation contractors do not guarantee sterility for the materials delivered to them.

1. They assist in developing and validating an acceptable dose for sterilisation. This is based on the bioburden (ISO, 1995) of the items offered for sterilisation.

2. They assist in determining a loading pattern of the client's containers within their irradiation carriers or containers, which ensures that the validated acceptable dose of radiation is delivered as a minimum to all parts of the items being treated.

3. They guarantee by means of process control records and supportive dosimetry that that the chosen acceptable dose is delivered to all parts of the materials delivered to them for irradiation, assuming that the presentation, loading, orientation of items within the client's containers are as they were during process development and validation.

The critical part of this is that the presentation of materials for irradiation within the client's transportation containers remains consistent. The contractor generally places dosimeters on the external surfaces of containers, not at the actual low dose point — his assurance of delivery of the minimum acceptable dose to the low dose point is based on knowledge gained from dose mapping of defined loads during dose development and validation. This information is valueless if the client does not maintain a consistent loading pattern.

Some pharmaceutical manufacturers using contract irradiators choose to 'keep tabs' on them by including dosimeters of the same type (Red Acrylic) within their product. However, this type of dosimeter is subject to quite a lot of variability, particularly to a phenomenon called 'fade' in which readings change according to the temperature of irradiation, the time between completion of radiation and read-out, and the temperature and humidity conditions between end of irradiation and read-out. These are all rather specific to the irradiator and to each batch of Red Acrylic dosimeters. As a consequence most irradiation contractors store their dosimeters under controlled conditions and read at a standard, but arbitrary, time after completion of irradiation, to allow comparison to a standard curve based on the same conditions. None of this is controllable for 'sneaky' dosimeters, so one might wonder if the value merits the cost.

Annex 1 of the EU GMPs (Anon, 2004(a)) mandates use of 'radiation sensitive colour disks' on each package of materials. These are normally yellow, turning red after irradiation. The second part of the clause in Annex 1 (Anon, 2004(a)) goes on to say that these disks 'differentiate between packages which have been subjected to irradiation and those which have not'. It is important to recognise that the colour change does not give any assurance that the validated acceptable dose has been received, only that the disks have been exposed to some form of irradiation.

Hydrogen Peroxide Vapour (HPV/VHP)

Vapourised hydrogen peroxide is normally used as a surface 'sterilant' in spatially enclosed environments where the combination of temperature, humidity and concentration can be carefully controlled to approach an acceptable level of effectiveness. As penetration/condensation of the hydrogen peroxide may be limited, most items will need to be thoroughly exposed to the vapour. When VHP is used in an isolator environment (its most common usage) items are cleaned and then directly exposed to VHP within the isolator enclosure. To ensure sanitisation or sterilisation (it is doubtful whether VHP under practical conditions can provide the same level of sterility assurance as the other treatments addressed above — for this reason there is a debate about whether it should be referred to as a sanitiser or sterilant) of items it is necessary to ensure that all surfaces are exposed and none are occluded. For the most part this is done in isolators by placing items on perforated shelves or in perforated trays, but in some instances critical items may be suspended from (say) hooks. Protective wrapping is counter-productive.

The only exception to this is environmental media plates, which must be sealed in plastic bags to prevent the vapourised hydrogen peroxide absorbing into the media and inhibiting its growth promotion capabilities.

PROTECTION AFTER STERILISATION

It is important that materials once sterilised, maintain their sterility up to their point of use (and indeed during use, but this is outside the scope of this chapter). In this context we have referred to protective containers and wrappings having the capacity to withstand the rigours of the chosen sterilisation process, and subsequent handling and transportation. It is sensible to remember that the sterilisation process may have altered the properties of the protective material, or of its sealing, and these factors should be taken into account when choosing and evaluating protective systems.

For instance, a wrapping material should easily fold away from a package to help ensure aseptic presentation in the Grade A/Class 100 area. The material should not snap back over a package once it is opened, since this can contaminate the contents. It is commonly observed that paper-based sterilisation wrap tends to hold the shape of the wrapped item, especially after sterilisation.

However, in addition to this there are other factors which need to be seriously considered.

The subject of dry heat tunnels for sterilisation and depyrogenation has already been addressed. Sterile pyrogen-free vials are delivered continuously to the filling line at the time and in the numbers required. This type of technology is the exception rather than the rule. Most sterilisation is done 'batch-wise'. A defined number of items (say rubber stoppers of the correct size for a projected batch of product) will be sterilised beforehand, as will a filling set up of pumps, manifolds and needles, etc. This invites questions concerning how long these items should be kept before use and under what conditions they should be kept.

For the most part, materials which are barriers to ingress of microorganisms remain so unless and until some event occurs to disrupt them. Their susceptibility to damage by disruptive events may increase over time, the polymeric structures of flexible wrappings may be affected by sterilisation processes with the result that they become more and more brittle (less and less flexible) over time after sterilisation. Papers are particularly susceptible to tearing after sterilisation, particularly if they are wet. Minimal handling is therefore recommended.

In modern designed aseptic sterile manufacturing facilities, it is common for double-ended sterilisers to open directly into Grade B/Class 10,000 (or better) aseptic core filling rooms. Pass-through hatches for irradiation sterilised items also open into the Grade B area so that what is needed is supplied when it is needed.

In older facility designs it is common to find 'cooling' or 'storage' rooms within the aseptic core. It is human nature that the availability of storage space encourages its utilisation, extra sterile materials are held 'to be sure to be sure'. The longer things are stored the greater their susceptibility to disruptive events, leading possibly to damage and eventually to contamination of the sterile contents. The same applies to extra handling — the more times and the greater distances over which items are transported, the more susceptible they are to disruptive events.

However, in reality many facilities have been designed with cooling rooms and cannot be re-designed without numerous 'knock-on' effects which make this impractical. For instance, most often dry heat oven through-put capabilities, and registered operating practices tend to be inextricably intertwined into facility design. Dry heat sterilisation cycles tend to take a very long time due to slow rates of heat transfer during heat up phases, without also having to add lengthy cool-down phases in the ovens. As another example, rubber stoppers sterilised in autoclaves are rarely usable straight from the autoclave — they are too hot, may have to undergo some humidity equilibration, and they may quite simply not run on the tracks until cooled off. For reasons of optimising throughput on ovens and autoclaves, older facility designs have tended to include a 'cooling area', and dry heat sterilisation processes have been registered without lengthy cool-down phases in the ovens.

Whichever option, cooling rooms or direct supply to the filling room, sterilised items are expected to be stored in Grade A/Class 100 conditions (ie under unidirectional HEPA filter air). However, unless the whole filling room can be graded/classified as Grade A/Class 100, it will be necessary to move stored sterilised materials through an area (perhaps only a metre or so, but possibly much larger distances) of lower classification. For compliance with both EU (Anon, 2004(a)) and US (FDA, 2004) GMPs this 'background' room may be no worse than Grade B/Class 10,000. Filling rooms with complete unidirectional air-flow ceilings which allow classification as Grade A/Class 100 are quite unusual in Europe, but less so in the US where energy costs have traditionally been much lower.

This transfer of sterilised materials poses a problem which impacts on the earlier processes of preparing and protecting materials for sterilisation. It has led to the extensive use of double bagging in purpose made paper, Tyvek® or similar bags or for large or awkwardly shaped items by double wrapping in paper or Tyvek®. Double wrapping is used so that after storage the 'contaminated' outer wrapping can be removed prior to exposing the item within the critical filling area. For steam sterilisation it is clear that the extra bags or wrap pose additional barriers to steam penetration/air removal and this must be factored into the validation process.

Over the last decade of the 20th century and first decade of the 21st, there has been regulatory pressure in the UK at least towards maintaining a continuity of Grade A/Class 100 conditions for all materials and equipment from the off-loading of double-door sterilisers to point of use. Manufacturers have reacted differently to this kind of pressure, but it is not unusual to find double containment systems used. Where storage is necessary within the aseptic suites this is carried out in cabinets provided with unidirectional HEPA-filtered air protection — sometimes even using mobile units.

Validation of storage times for sterilised items and components is usually done through media fills. The equipment and components to be used here are generally prepared as long as is permitted before the planned 'run', before sterilisation, after sterilisation, and then set up for the media fill.

SUMMARY

Any item used within the sterile core of aseptic pharmaceutical manufacturing facilities, whether direct or indirect product or non-product contact, must be presented at worst with a low bioburden, and at best sterile and non-pyrogenic. For inert, robust items the actual sterilisation poses few problems, but due to the batch process used for aseptically prepared parenterals, most items will be prepared in advance and some period of storage will be required.

To ensure that sterilised products remain sterile up to the point of use some form of post-sterilisation protection is required. This can take a variety of forms — from rigid stainless steel containers through paper to Tyvek® wrappings. Manufacturers need to consider carefully the vagaries of the sterilisation process and what would constitute a barrier to the effectiveness of the chosen method of protection. They also need to look at what impact the sterilisation method has on the chosen barrier material, both in the short and long term, where applicable.

Finally, a pragmatic approach needs to be taken to cleaning the items for sterilisation. This should be implemented only after a careful consideration of the risks involved, rather than performance by rote, or in 'blind' accordance to supposed regulatory expectations.

REFERENCES

Anon (1997) European Standard EN 868: *Packaging materials and systems for medical devices that are to be sterilised.*

Anon (1999) Monograph No 10 — Cleaning Validation. Pharmaceutical Quality Group, Institute of Quality Assurance, UK

Anon (2004(a)) *The Rules Governing Medicinal Products in the European Community. Volume IV. Good Manufacturing Practice for Medicinal Products*. Annex 1 Manufacture of Sterile Medicinal Products

Anon (2004(b)) *The Rules Governing Medicinal Products in the European Community. Volume IV. Good Manufacturing Practice for Medicinal Products*. Annex 2 Manufacture of Biological Medicinal Products for Human Use.

Anon (2004(c)) *The Rules Governing Medicinal Products in the European Community. Volume IV. Good Manufacturing Practice for Medicinal Products*. Annex 15 Qualification and Validation.

du Moulin, D.C., Doyle, G.O., MacKay, J. and Hedley-Whyte, J. (1981) Bacterial fouling of a hospital closed-loop cooling system by *Pseudomonas* sp. *Journal of Clinical Microbiology* **13** (6) 1060–1065.

EMEA (2002) Committee for Proprietary Medicinal Products/Committee for Veterinary Medicinal Products. *Note for guidance on the quality of water for pharmaceutical use*. EMEA, London

FDA (1993) *Guide to Inspections of Validation of Cleaning Processes* Department of Health and Human Services, US Food and Drug Administration.

FDA (2004) *Guidance for Industry: Sterile Drug Products Produced by Aseptic Processing — Current Good Manufacturing Practice*, Department of Health and Human Services, US Food and Drug Administration, CDER, CBER, ORA, September 2004.

Halls, N.A. (2004) Media fills and their applications. In *Microbiological Contamination Control in Pharmaceutical CleanRrooms* ed N.A. Halls. Boca Raton: CRC Press.

Hamilton, W.B. (1931) Gum production by *Azotobacter chroococum* of Beijerinck and its composition. *Journal of Bacteriology* **22** (4) 249–254.

International Standards Organisation (ISO) (1995) IS 11137 *Sterilisation of health care products. Requirements for validation and routine control — radiation sterilisation.*

International Standards Organisation (1997) IS 11607: *Packaging for terminally sterilised medical devices.*

Le Blanc, D.A. (2005(a)) Dispelling cleaning validation myths: part 1. Pharmaceutical Technology Europe, November 2005.

Le Blanc, D.A. (2005(b)) Dispelling cleaning validation myths: part 1. Pharmaceutical Technology Europe, December 2005.

Parenteral Drug Association (1996) Technical Report No 24. Current practices in the validation of aseptic processing. *PDA Journal of Science and Technology* **51**: Supplement S2

Riedewald, F. (2006) Bacterial adhesion to surfaces: the influence of surface roughness. *PDA Journal of Pharmaceutical Science and Technology* **60** (3) 164–171.

Shimomora, Y. , Ohno, R., Kawai, F. & Kimbara, K. (2006) Method for assessment and morphological changes of bacteria in the early stages of colony formation on a simulated natural environment. *Applied and Environmental Microbiology* **72** (7) 5037–5042.

Sinclair C.S, and Tallentire A, (1981) Penetration of Medical Grade Papers by Airborne Bacterial Spores, *Journal of Pharmacy and Pharmacology* **38**:77.

Tallentire A, and Sinclair C.S (1985), Microbiological Barrier Properties of Uncoated and Coated Spunbonded Polyolefin (Tyvek) *Medical Devices and Diagnostics Industry* **7**(10): 5054.

AUTHOR BIOGRAPHIES

Dr Nigel Halls is Executive Director for Science and Technology and a founding partner of IAGT (International Academy of GMP Training) Ltd — one of Europe's fastest growing training and consultancy organisations.

He has degrees in Microbiology from the Universities of Bradford and Bath (UK) and has worked in the pharmaceutical and sterile medical devices industry for many years. His principle professional interests are sterile manufacture, manufacture of aqueous pharmaceutical products, and design, control and monitoring of pharmaceutical water systems.

Dr Halls' extensive publications include *Achieving Sterility* (1994) and *Microbiological Contamination Control in Pharmaceutical Clean Rooms* (2004). He recently scripted multi-media pharmaceutical training programmes for MVI International. He serves as member of the Fellowship and Membership Committee of the Institute of Biology (UK) and is an active member of PDA. At the time of writing he is serving as a member of the Special Task Group charged with revising Technical Report Number 1 — Validation of Moist Heat Sterilization Processes: Cycle Design, Development, Qualification and Ongoing Control.

Dr Halls relaxes by walking his dogs and listening to opera. He lives by the sea on the south coast of England.

Lucia Clontz is Director of Microbiology and Chairman of the Operational Excellence Team at Diosynth Biotechnology, RTP, North Carolina, USA. She has over 20 years experience in the pharmaceutical and biotechnology industries. In her previous position as a Consultant and Director of Regulatory Compliance at Serentec, Inc., Lucia helped clients find effective and practical solutions for microbiology and microbial contamination issues associated with facilities, processes, and laboratories. Lucia received a B.S. degree in Chemistry (with concentration in Chemical Engineering) from the Federal University of Rio de Janeiro, Brazil and a B.S. in Marine Sciences (*Magna Cum Laude*) from the University of South Carolina, SC, USA. Lucia also holds a Master of Science degree in Microbial Biotechnology and a Certificate in Molecular Biotechnology from the North Carolina State University, NC, USA. She is a member of the PDA, co-founder of the Pharmaceutical Microbiology Forum (PMF) Organization, teaches courses worldwide, and is a published author.

Simon Cole completed his university education in Microbiology in 1982. He has worked in the filtration industry for 24 years, managing technical support functions in two major international filtration companies and specialising in supporting filtration applications for the pharmaceutical and biotechnology industries. He has travelled worldwide to perform practical evaluations and to provide advice and training in all aspects of filtration. Simon has established validation support services to enable clients to achieve regulatory compliance for sterilising filtration systems. He has published articles in a number of journals and spoken at international conferences on aseptic processing and filtration technology, for diverse industrial applications ranging from water purification and drug sterilization to blood fractionation and virus reduction in biopharmaceuticals.

Simon has also recently written a chapter concerning steam sterilisation of filtration systems for the 2nd edition of *Filtration and Purification in the Biopharmaceutical Industry* (Eds. Jornitz M & Meltzer T), to be published in 2007, and is a member of the Special Interest Group on Aseptic Process Filtration for the Pharmaceutical and Healthcare Sciences Society (formerly the Parenteral Society).

Currently living in France, Simon is focussing on bilingual technical training and support services for the biopharmaceutical industry, covering filtration as well as primary and secondary packaging operations. These services are combined with technical interpretation and related language services.

Tim Coles has been active in the field of pharmaceutical isolation technology for over 20 years. He has worked with several isolator manufacturers, including his own company, Cambridge Isolation Technology. Tim has published a major work on isolation (*Isolation Technology — a Practical Guide*. CRC Press Inc., 2004), is a co-editor of the *Yellow Guide* (Pharmaceutical Isolators, Pharmaceutical Press, 2004) for isolator operation and has contributed to the forthcoming PDA monograph on biological indicators for sporicidal gassing. He has BSc and MPhil Degrees in Environmental Sciences and is currently a director, together with his wife Caroline, of Pharminox Isolation Ltd.

Stewart Green trained as a microbiologist and also has a degree in applied biology, and a Masters in Strategic Quality Management. He has occupied senior roles in manufacturing, validation, QC, QA, and logistics in a 40-year career in the pharma industry. Currently he is Director of Quality for Wyeth with a reporting group of 120 people covering a multitude of disciplines. He has a career-long interest in aseptic manufacture and has contributed to a number of books on the topic.

Kevin Williams is Associate Pharmacological Consultant, Eli Lilly & Co., Indianapolis, Indiana. The author or co-author of several books and peer-reviewed articles, Mr Williams is a member of ASM International and the Parenteral Drug Association. He received the BS degree (1982) from Texas A&M University, College Station, USA.

INDEX